GOLD DEPOSITS:

Origin, Evolution, and Present Characteristics

An analysis of the present gold deposits in terms of the geometrical, mechanical, thermal, and chemical aspects of the Earth's behavior during the past 4.6 billion years.

J. H. TATSCH

TATSCH ASSOCIATES
Sudbury, Massachusetts 01776
1975

Printed in the United States of America, September, 1975. Copies are available from Tatsch Associates and from book dealers in all parts of the world.

International Standard Book Number: 0-912890-07-X.

Library of Congress Catalog Card Number: 75-1947.

Preface

Gold deposits have formed in all geologic epochs from the earliest Archean to the recent Cenozoic. These deposits developed at different stages in the evolution of mobile belts, platforms, and stabilized structures of the Earth. The long-lived, selective ubiquity of gold deposits has always fascinated mankind. This book takes a look at some of the intriguing questions about where gold came from and why the deposits formed at some places but not at others.

For millions of years, man has gone into the mountains, in all parts of the world, to look for gold. Many times, he has found gold there. He has found it also, sometimes, on the plains and in the rivers that have eroded from the mountains. To the itinerant prospector, this meant that "thar's gold in them thar hills." To the more-circumspect observer, it meant that there must be some relationship between gold deposits and the mountains. But no one really knows what correlation there might be between gold deposits and the mountains.

This book presents one possible answer to this relationship. It hypothesizes that the origin, evolution, and present characteristics of gold deposits are related to the same mechanism that produces the mountains. This mechanism is the "tectonospheric Earth model," which has been described in the book, *The Earth's Tectonosphere,* and used as the basis for the books, *Mineral Deposits* and *Petroleum Deposits,* by the same author.

One of the strengths of the Tectonospheric Earth Model is that it permits the prediction of certain variations that have occurred within the geometrical, mechanical, thermal, and chemical heterogeneities of the Earth's upper 1000 kilometers during the past 4.6 billion years. The ability of the model to provide these predictions forms the basis for its use as a supplementary tool in the exploration for gold deposits.

Table of Contents

1. Gold Deposits: A Global Survey 1
2. The Tectonospheric Earth Model: A New Concept 10
3. The Geometrical, Mechanical, Thermal, and Chemical Behavior of the Earth during the Past 4.6 Billion Years: A Summary 36
4. Plate Tectonics, Omniductive Processes, and Seismotectonomagmatic Belts 44
5. Intrusive and Extrusive Activity: A Long-Lived, Deep-Seated Basis for the Origin and Evolution of Mineral Deposits 51
6. Cenozoic Seismotectonomagmatic Belts and the Associated Gold Deposits 73
7. Mesozoic Seismotectonomagmatic Belts and the Associated Gold Deposits 118
8. Paleozoic Seismotectonomagmatic Belts and the Associated Gold Deposits 143
9. Proterozoic Seismotectonomagmatic Belts and the Associated Gold Deposits 165
10. Archean Seismotectonomagmatic Belts and the Associated Gold Deposits 194
11. The Present Distribution of Gold Deposits as a Function of the Earth's Behavior during the Past 4.6 Billion Years 206
12. The Tectonospheric Earth Model as a Supplementary Tool in the Exploration for Gold Deposits 214

References 218

Index 260

Chapter 1

GOLD DEPOSITS: A GLOBAL SURVEY

The Earth's gold deposits were formed in all geologic epochs, from the Archean through the Cenozoic, at different stages of development within mobile belts, platforms, and stabilized structures. Magmatic melts and transmagmatic solutions of various compositions have served as a source of the gold mineralizations. Today we see commercial gold deposits of three general types (magmatic, hydrothermal, and metamorphogenetic) plus a large number of modifications and combinations of these types. Essentially all types of large geotectonic structures, at various stages of their formation, are characterized by gold mineralizations.

The earliest gold deposits, associated with the Archean, are most obvious in the shields of Africa, Asia, Australia, and the Americas. Proterozoic gold deposits formed within platforms and their folded peripheries, including those of eastern Siberia, southern and western Africa, the Americas, and Australia. Paleozoic gold deposits formed primarily in western Siberia, in eastern Sayan, in China, and in Canada. Mesozoic gold deposits formed primarily in the Urals, in Kazakhstan, in central Asia, in North America, and in Australia. Cenozoic gold deposits formed primarily along the Pacific "ring of fire" and along the Mediterranean-Indonesian belt.

Gold Deposits and Geotectonic Structures.

It has long been known that various mineral deposits are closely associated with different geotectonic structures. More specifically, it appears that the conditions of formation and the regularities governing the distribution of gold deposits can be clearly identified with a panorama of seismic, tectonic, and magmatic activities that have occurred within the Earth during the

4.6 b. y. that the Earth is believed to have been in existence. It is well to review some of the structures that are seen today in association with gold deposits as vestiges of the Earth's past behavior:

a. The Earth's continental shields are characterized by the association of gold mineralization with narrow synclinal zones of the sutural type filled with predominantly volcanogenic green rocks. These deposits are related paragenetically with dikes, stocks, and some larger masses of granitoid rocks. Thus, the gold deposits of the Canadian shield are represented mainly by quartz lodes, by zones of sulfide impregnation, and by auriferous sulfide deposits of the pyrite type, including chalcopyrite, sphalerite, pyrrhotite, arsenopyrite, and tellurides. A characteristic feature of the African-shield areas is the abundance of contiguous deposits of small size, represented primarily by about 5000 quartz lodes. The quartz lodes of the Kolar deposits of the Indian shield are associated with a narrow syncline comprising amphibolites of the Dharvar series, with vein ores represented by scheelite, magnetite, ilmenite, pyrite, galena, pyrrhotite, arsenopyrite, chalcopyrite, native gold, gold tellurides, and silver. The Baltic shield contains small deposits of gold in quartz veins and in association with pyrite ores.

b. The Proterozoic structural stages of old platforms are characterized by gold deposits of metamorphosed auriferous conglomerates, including those in the Witwatersrand and Tarkwa mines of Africa. Similar deposits are found in the Bland-River area of the southern part of the Canadian shield, where Proterozoic uraniferous conglomerates contain small concentrations of gold. Similarly, the gold deposits of Ghana are represented by quartz lodes and vein zones, associated with deep faults along the boundaries of synclinal depressions containing auriferous conglomerates that apparently were intruded along the deep faults. The Homestake deposits of the Canadian shield also are associated with an old platform.

c. In some cases, platforms have been modified by old folded belts, where complex intrusions of basic magmas produced auriferous Cu-Ni-Pt ores. Examples of this type include those of Sudbury, Bushveld, and Norilsk.

d. Within miogeosynclinal depressions, gold ores are formed in association with geosynclinal granitoids. These include the Lena and Enisey gold deposits within the upper Proterozoic of Siberia; the Victorian deposits within the Paleozoic of Australia; and the South Verkhoyan, Amur, Yanokolyma, and Chukotka deposits of the Mesozoic.

e. The most diverse gold deposits are found within the eugeosynclinal areas of volcanogenic belts. During the geosynclinal stage, numerous pyrite deposits of auriferous ores were formed in association with basaltic volcanism. Most of the gold is deposited contemporaneously with the early pyrites. The gold of this stage is finely dispersed. Later gold deposits in these formations are associated with the deposition of sulfides of the chalcopyrite-sphalerite type. Sometimes, a superimposed gold mineralization is associated with later orogenic and post-orogenic stages of volcanogenic-belt development. Differences within geosynclinal gold deposits may be seen by considering the pre-orogenic, orogenic, and post-orogenic deposits separately:

(1). The <u>pre-orogenic</u> gold deposits are represented by quartz lodes, often associated genetically with hypabyssal intrusions of granodiorite and granosyenite formed during the final stages of basaltoid magmatism. Some of these intrusions are associated with gold-bearing skarns, including deposits of the southern and middle Urals, in western Siberia, in central Asia, and in similar systems of other continents.

(2). The orogenic gold deposits are genetically related to granitoid intrusions formed during episodes of inverse uplift and immediately following or accompanying folded dislocations. Some of these are related to batholithic intrusions. No major geological events separate the intrusions and the deposition of the gold ores. However, the formation of these deposits is usually preceded by minor intrusions and dikes and by cross-cutting granitoidal batholiths that are comagmatically related to the deposits. Examples of these include the deposits of Kochkar, Getygara, Kuznetsk, and Altau.

(3). The post-orogenic gold deposits are closely associated with belts of dikes and minor intrusions comprising granitoid rocks and lamprophyres. Examples include the deposits of northern Kazakhstan, the Altai-Sayny folded areas, the middle Urals (e. g., Beresovsk), and those from similar structures in other continents.

f. Gold deposits within areas of tectonic activity of consolidated structure fall into two types of auriferous provinces:

(1). Gold mineralizations within old platforms of tectonic activity are paragenetically associated with the formation of alkaline-earth and alkaline minor intrusions. These are localized mainly in the Cambrian carbonaceous series with some in the Jurassic terrigenous deposits of the upper structural stage overlying an Archean crystalline basement. The intrusions appear to be controlled by deep faults connected with flat platforms and similar structures.

(2). Gold mineralizations within areas of tectonic activity with completed folding are paragenetically associated with the formation of granitoid intrusions occurring as chains in the axial parts of arched uplifts. Most of these intrusions comprise Mesozoic magmatism superimposed on orogenic activity. Compositionally, the intrusions correspond to the gabbro-dioritic-granitoid type. Later associated minor intrusives comprise differentiated series of rocks from basaltoid to granitoid compositions. Jurassic gold deposits are closest related, spatially and temporally, to these formations. Examples include the quartz-reef deposits at Darasun and the stockworks at Klyuchi.

g. The latest stages of Mesozoic tectonic activity is characterized by the formation of near-surface chalcedony-quartz lode deposits of gold-silver. The magmatism is sub-aerial and comprises rocks mainly of the liparite series with a fairly high alkalinity. Typical deposits are represented by quartz-lode bodies, multiple veins, and stockworks (e. g., the Balei type found in most continents).

h. Gold deposits associated with recent volcanogenic belts comprise near-surface gold-silver formations. These are found in most continents, normally characterized by a possibly paragenetic spatial association with subvolcanic intrusions of liparites. These are formed at the final stages of the volcanogenic-belt development and have a more mixed composition than do the effusives of these belts. Some auriferous deposits occur near the centers of volcanic activity. The ore bodies comprise banded chalcedony comb quartz, adularia, and carbonates. Examples include Cripple-Creek, Homestake, Goldfields, and deposits in similar formations of most continents, including those within the Okhotsk-Chukotka link to the Asian volcanogenic belt and those within certain parts of the Koryako-Kamchatka folded area.

Gold Formations.

There is no standardized method for classifying gold formations. However, for the purpose of this book, it is well to classify gold deposits in terms of the volcanogenic environments in which they have developed. When this method of classification is used, there are 10 types of gold formations:

a. Gold-quartz. These formations are associated with the Earth's earliest, complexly-folded Archean and early Proterozoic structures. They appear to be related to granitoidal intrusions of the batholithic type. Biotization, chloritization, and tourmalinization have changed these formations into veins and vein systems of auriferous deposits.

b. Gold sulfides with copper sulfides and platinoids. These formations are associated with the upper stages of volcanogenic-belt development, particularly where the belts are sutured to the marginal parts of the platforms (See, e. g., Tatsch, 1972a: chapter 7; 1973a). They appear to be related to composite differentiated intrusions of basic rocks. Skarnization, post-skarn metasomatism, and transformation have changed these formations into deposits and zones of impregnated rocks.

c. Gold sulfides with copper pyrites and polymetallic pyrites. These formations are associated with eugeosynclinal structures of the pre-orogenic stage of volcanogenic-belt development.

They appear to be related to derivatiees of basaltic volcanicity. Silicification, sericitization, and chloritization have changed these formations into veins, bonanzas in skarns, and stockworks.

d. Gold-quartz with gold-quartz sulfides, and gold-copper. These formations are associated with eugeosynclinal structures of the orogenic stages of volcanogenic-belt development. They appear to be related to granitoidal intrusions. Biotization, silicification, sericitization, and chloritization have changed these formations into veins, bonanzas in skarns, and stockworks.

e. Gold-quartz with gold-quartz sulfides in geosynclines. These formations are associated with eugeosynclinal structures of the post-orogenic stages of volcanogenic-belt development. They appear to be related to minor intrusions and dikes mainly of the granodiorite series. Beresitization, listvenitization, and silicification have changed these formations into stockworks in dikes, veins, and systems of mineralization zones.

f. Gold-quartz with gold sulfides. These formations are associated with miogeosynclinal structures of the orogenic stages of volcanogenic-belt development. They appear to be related to granitic intrusions. Sericitization, chloritization, and silicification have changed these formations into veins and bonanzas.

g. Gold-quartz sulfides with gold-quartz. These formations are associated with miogeosynclinal structures of the post-orogenic stages of volcanogenic-belt development. They appear to be related to minor intrusions and dikes of granitoid and subalkaline series. Beresitization, chloritization, and silicification have changed these formations into stockworks in dikes, veins, vein systems, and mineralization zones.

h. Gold-quartz with gold-quartz sulfides in activated areas. These formations are associated with zones of platform activity within activated areas of volcanogenic-belt development. They appear to be related to minor intrusions and dikes of alkaline and subalkaline series. Ankeritization, dolomitization, anthophylitization, and silicification have changed these formations into deposits, veins, and bonanzas.

i. Gold-quartz sulfides with gold-quartz and gold-silver. These formations are associated with zones of activity in areas of final folding of volcanogenic-belt development. They appear to be related to (1) minor intrusions and dikes of granodiorite series, and (2) volcanic pipes and dikes of subvolcanic facies of acid composition. Tourmalinization, albitization, silicification, and beresitation have changed the minor intrusions and associated dikes into vein systems, stockworks, and veins. Aularization, kaolinization, and alunitization have changed the volcanic pipes and associated dikes into similar deposits.

j. Gold-silver with silver-gold. These formations are associated with intra and post-orogenic structures within areas of young eruptives in active volcanogenic belts. They appear to be related to volcanic and volcano-plutonic complexes. Kaolinization, alunitization, and zeolitization have changed these formations into veins, vein systems, and stockworks.

From the above 10 types of gold deposits, it may be seen that the conditions of formation and the regularities governing the distribution of gold deposits depends on their association with different geotectonic structures of the volcanogenic belts that have existed in various parts of the Earth during the 4.6 b. y. that the Earth is believed to have been in existence.

Closer scrutiny of these 10 types of gold formations suggests that they are only 10 of a continuum of types of gold formations that have developed during the past 4.6 b. y. This observation is used in the following chapters to show how a "tectonospheric Earth model" may be used to explain the conditions of formation and the regularities governing the distribution of gold deposits as a function of their association with different geotectonic structures that have existed in various parts of the Earth.

Chapter 2

THE TECTONOPSHERIC EARTH MODEL: A NEW CONCEPT

One of the theses of this book is that the Tectonospheric Earth Model (Tatsch, 1972a) may be used as a supplementary tool in the exploration for gold deposits. The reader not familiar with this model should consult the cited text for the geometrical, mechanical, thermal, and chemical details regarding the model. For the reader already familiar with the model, this chapter is intended to serve as a review.

The Basic 5400-km Primordial Earth.

When a dual primeval planet hypothesis is applied to an analysis of the early solar system, one of the consequences is a primordial Earth and a co-orbital companion, Earth Prime. Each of these modified Bullen-type bodies had a radius of about 5400 km and was fractured into octants early in the history of the solar system. The 5400-km primordial Earth became the "subtectonosphere" of the present Earth. The Earth's tectonosphere (the upper 1000 km in this model) is an accretion onto the primordial Earth of the remnant substance of five of the octants of Earth Prime, which had separated into octantal parts early in the history of the solar system. The early Earth may be represented schematically as shown in Fig. 2-1.

The center, C, represents a solid core, with an undefined radius assumed to lie somewhere within the range 0 to several hundred km. The hatched spherical shell, MC, represents a molten core. The 8 octants of the primordial mantle surround the molten core and are marked S. The preferential flow of heat, and possibly of volatiles, outward from the fracture belts, F, causes igneous activity at I.

The primordial Earth has a volume such that, when 5/8 of an

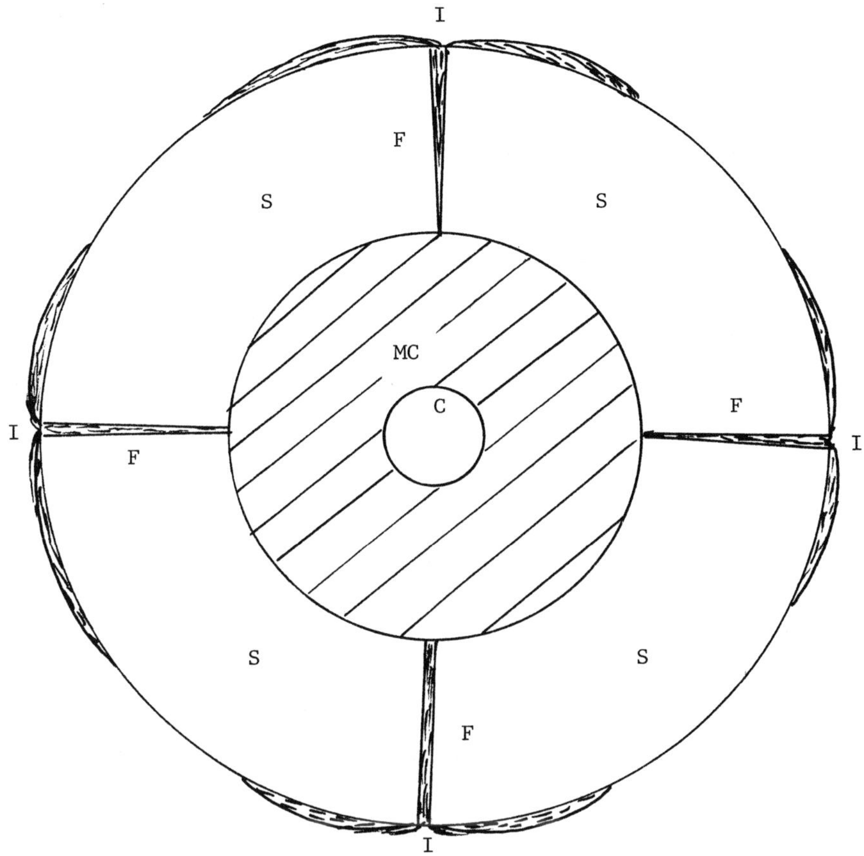

Figure 2-1. The 5400-km primordial Earth which, according to the Tectonospheric Earth Model, has been cut by three mutually orthogonal planes passing through the center. Schematic; not to scale. See text for details.

equivalent volume is added, the resulting body has a radius of about 6371 km, the radius of the actual Earth. This gives the primordial Earth a radius of about 5400 km, thus:

1 5/8 V' = V, where V' and V are the volumes of the primordial Earth and of the actual Earth, respectively. Then, since

$$V = 4/3 \text{ pi } R^3, \text{ we may write}$$

$$1.625\ (R')^3 = R^3, \text{ from which}$$

$$R' = (1.625)^{-1/3}\ R$$

$$R' = (1.625)^{-1/3}\ (6371 \text{ km}) \doteq 5400 \text{ km.}$$

This is the radius of the primordial Earth, as well as of Earth Prime and of the subtectonosphere used in the model.

The thickness of the tectonosphere is about 6371 km minus 5400, equals 971, or about 1000 km, roughly. Although the values, 5400 km and 1000 km, are used in this model for the radius of the subtectonosphere and for the thickness of the tectonosphere, respectively, it should be pointed out that these values are approximations which depend somewhat upon the particular assumptions made in connection with the assignment of initial values to the variable parameters of the basic tectonospheric Earth model derived from the dual primeval planet hypothesis.

Since the primordial Earth is fractured into octants by 3 mutually-orthogonal central planes, its surface is encircled by 3 mutually-orthogonal great-circular fracture belts, each having a length of 2 pi times 5400 km = 34,000 km, roughly, or about 3 x 34,000 km = 102,000 km total for the three belts.

The Geometrico-Mechanical Behavior of the Basic 5400-km Primordial Earth.

In order to analyze the geometrico-mechanical behavior of the basic 5400-km primordial Earth, it will be assumed that the basic behavior of the primordial Earth is such that it tends to equilibrate itself to a state of minimum energy. The driving mechanism of the system may be expressed as a function of the disequilibration energy inherent in its initial state. In simplest terms, it consists essentially of the resultant of two factors, each of which has been varying in a predictable manner during the past 4.6 b. y.: (1) the potential energy inherent in the geogenetically disequilibrated shape of the basic 5400-km primordial Earth; and (2) the selectively-channeled energy from the preferential flow of heat, and possibly of volatiles, outward from the three mutually-orthogonal belts which are "active" along the tensile portions of the surficial traces of the 3 planes separating the solid portions of the primordial Earth into 8 octantal subtectonospheric blocks.

When the driving mechanism of the basic model is applied to the primordial Earth, the behavior will depend upon certain geometrico-mechanical constraints inherent in the model. The simpler of these constraints may be summarized in terms of several types of equilibration, or tendency to assume a status of minimum energy:

Rotary Equilibration. Fig. 2-2 shows schematically the expected basic behavior of the original 5400-km primordial Earth in response to the fundamental driving mechanism of the model from the standpoint of rotary equilibration. Basically, there are three degrees of internal rotational freedom, constrained by the configuration of the model to act along the three mutually-orthogonal planes along which the octantal fracturing occurred.

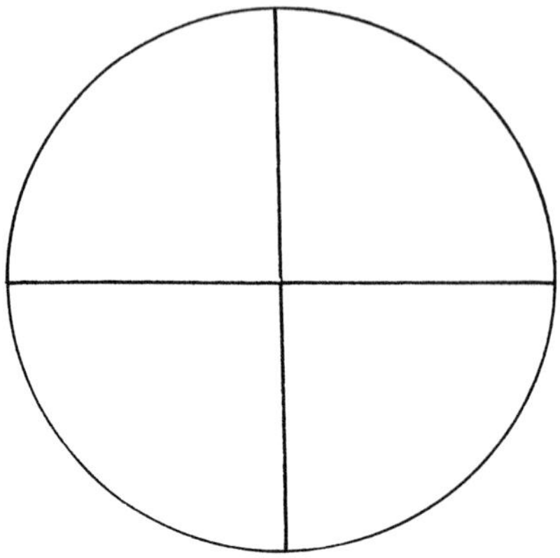

A. No rotation.

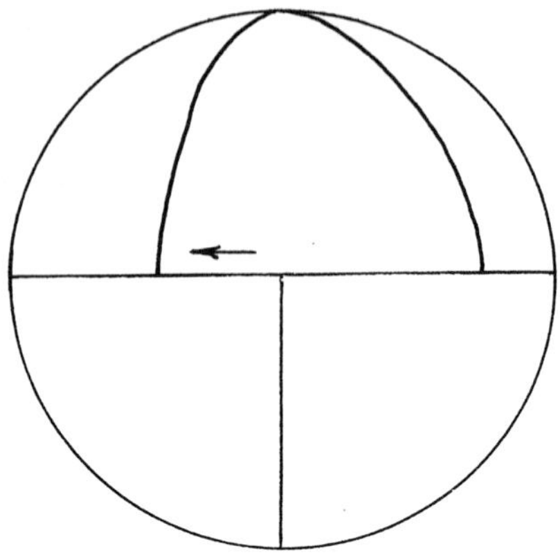

B. Rotation along the geophysical equatorial plane.

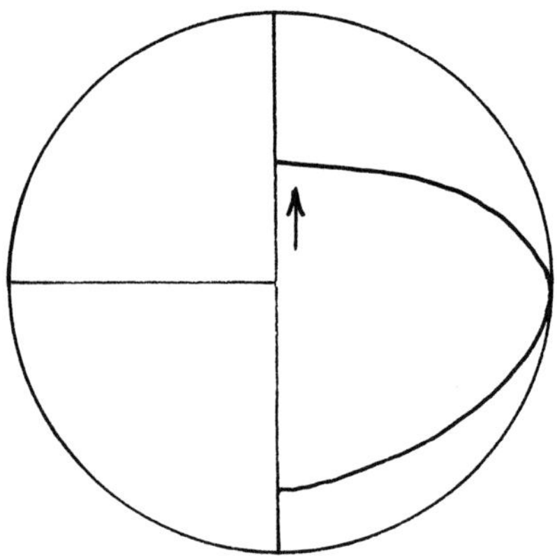

C. Rotation along the geophysical prime meridian plane.

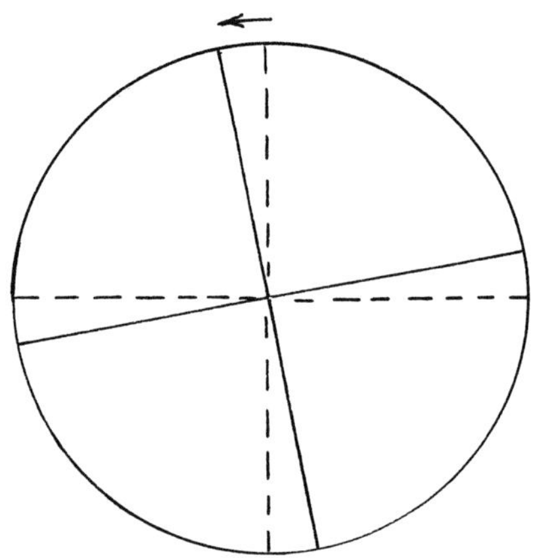

D. Rotation along the geophysical orthogonal meridian plane.

Figure 2-2. The expected basic behavior of the original 5400-km primordial Earth in response to the fundamental driving mechanism of the Tectonospheric Earth Model from the standpoint of rotary equilibration. Schematic; not to scale. See text for details.

It is noticed that, once the rotational motion has begun in any one of the three mutually-orthogonal fracture planes, the gross rotational motion in the other two planes is restricted until a net rotation of n pi/2 shall have transpired in the first plane, where n = 0, 1, 2, 3, - - -.

Radial Translatory Equilibration. Fig. 2-3 shows schematically the expected basic behavior of the original 5400-km primordial Earth in response to the fundamental driving mechanism of the model from the standpoint of radial translatory equilibration. The spherical octants are free to move radially with respect to the center of the model. From the principle of hydrostatic adjustment applied on a global basis, about half of the octants are subsided, at any one time, while the others are elevated with respect to the basic 5400-km radius. The magnitude of the radial motion visualized here is of the order of cm/yr, with maximum amplitudes of the order of ten km.

Transverse Translatory Equilibration. Fig. 2-4 shows schematically the expected basic behavior of the original 5400-km primordial Earth in response to the fundamental driving mechanism of the model from the standpoint of transverse translatory equilibration. Within the constraints of the model, some of the octants (normally two or four) may move transversally along the three mutually-orthogonal fracture-planes. The magnitude of this transverse motion is of the order of cm/yr, with maximum amplitude of the order of km.

Individual Octantal Motion. Fig. 2-5 shows schematically the expected basic independent behavior of the individual octants of the original 5400-km primordial Earth. Within the constraints of the model as a whole, each of the spherical octants

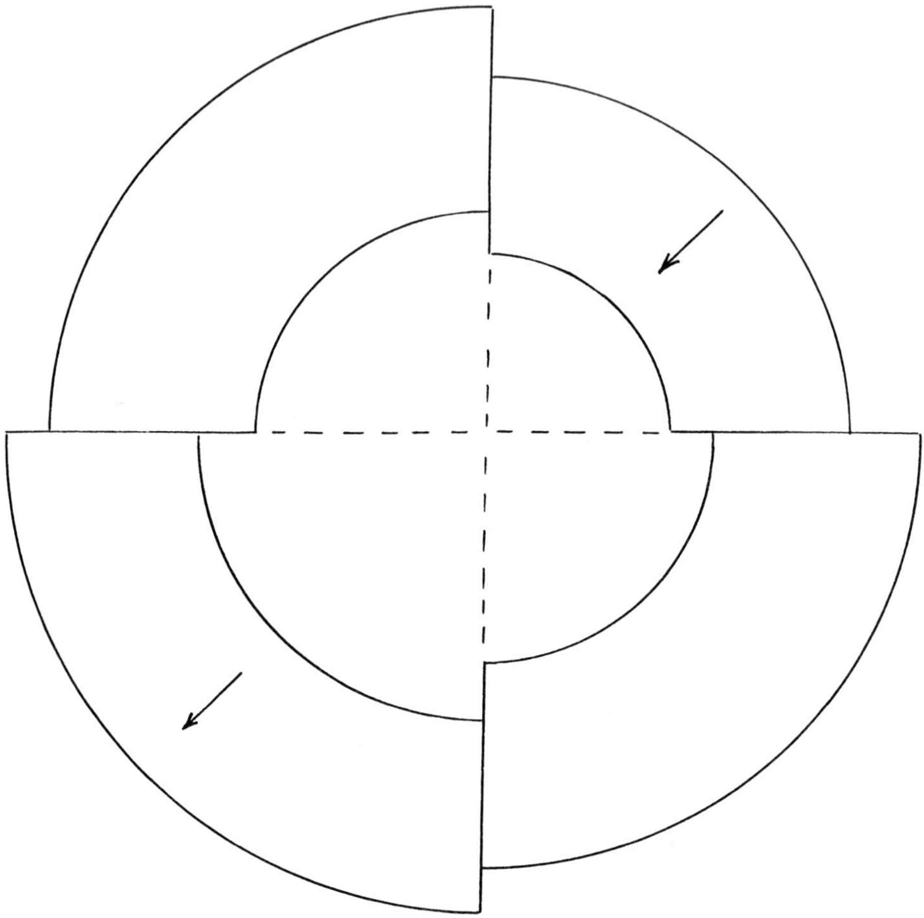

Figure 2-3. The expected basic behavior of the original 5400-km primordial Earth in response to the fundamental driving mechanism of the Tectonospheric Earth Model from the standpoint of radial translatory equilibration. Schematic; not to scale. See text for details.

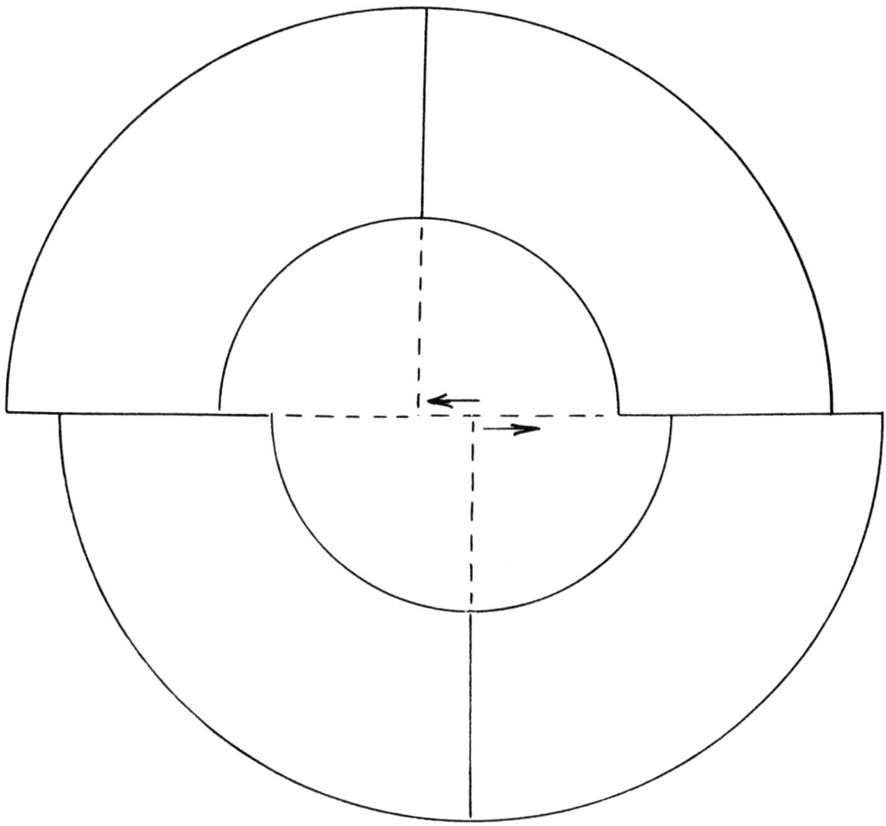

Figure 2-4. The expected basic behavior of the original 5400-km promordial Earth in response to the fundamental driving mechanism of the Tectonospheric Earth Model from the standpoint of transverse translatory equilibration. Schematic; not to scale. See text for details.

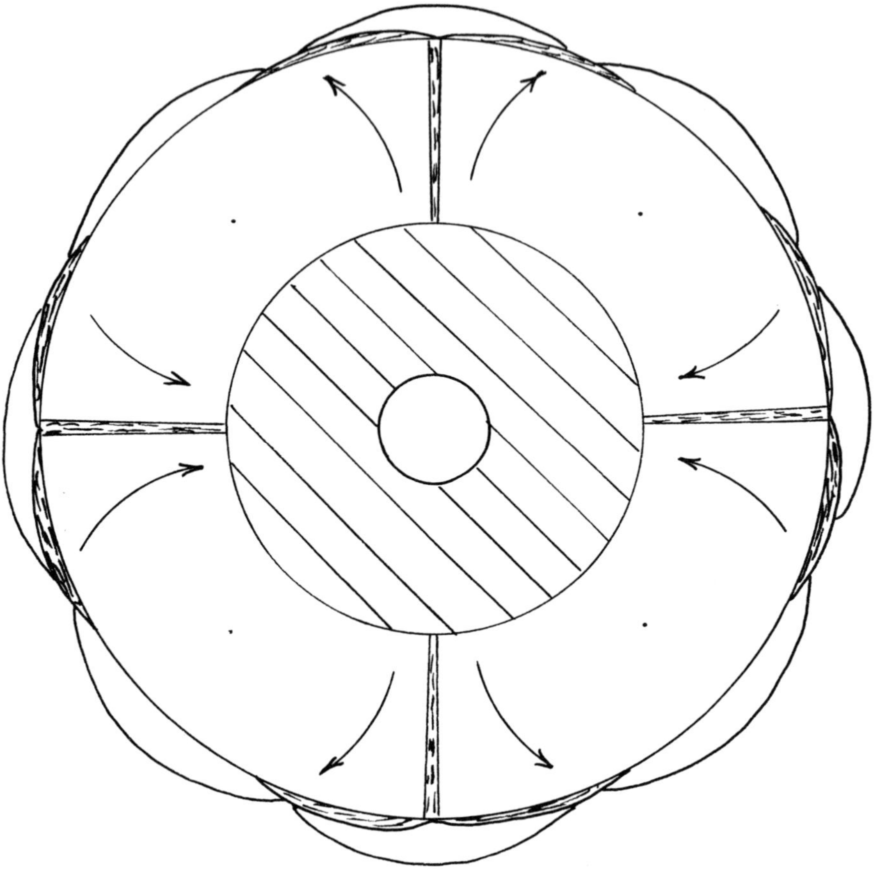

Figure 2-5. The expected basic <u>independent</u> behavior of the individual octants of the original 5400-km primordial Earth in response to the fundamental driving mechanism of the Tectonospheric Earth Model. Schematic; not to scale. See text for details.

has several degrees of freedom with respect to its individual center of gravity. Because of the constraints of the model, this evidences itself mainly as alternate tension and compression along the three mutually-orthogonal fracture planes. This effect is oscillatory, with a long time constant. At any particular time, other things being equal, tension should be found at about half of the octantal contact surfaces with compression at the others. Thus, since each great circle of the basic 5400-km primordial Earth is about 2 pi 5400 = 34000 km long, or 3 x 34,000 = 102,000 km total for all three great circles, then about ½ x 102,000 = 51,000 km of the surficial fracture pattern of the primordial Earth should be in tension while an equal amount is in compression (or neutral). Since the model is continually in the process of equilibrating itself, the locations of the specific segments which are in tension will tend to drift continually along the 102,000-km length of fracture system on the surface of the primordial Earth. Thus, from the standpoint of an external observer, the specific segments in tension would appear to be moving along the 102,000-km linear system in a cyclical but aperiodic manner, since the equilibration of the primordial Earth would, of necessity, be proceeding in a like manner (i. e., cyclically but aperiodically). The rate of such shifting of the tensile segments is of the order of cm/yr.

Segments not in tension at any given time are assumed to be in compression or in a neurtal state. Such segments, whether compressive or neutral, would lie between successive tensile segments and would, therefore, also shift along the 102,000-km subtectonospheric fracture system in a cyclical but aperiodic manner.

<u>Post-Primordial Tectonospheric Accretion</u>.

In evolving the post-primordial Earth from the basic 5400-km primordial Earth, three basic evolutionary modes may be assumed, depending upon the degree to which the 5 terrestrial octants of Earth Prime are fragmentized prior to accretion onto the primordial Earth. These modes are: (1) an octantal-fragment mode; (2) a multiple-fragment mode; and (3) a composite-fragment mode.

The Octantal-Fragment Accretionary Mode.

In the octantal-fragment mode, the 5 octants of Earth Prime are not completely fragmentized prior to accretion onto the basic 5400-km primordial Earth. In this case, parts of each of the 5 octants of Earth Prime might still exist today as unmodified portions of the Earth's tectonosphere, perhaps even as subsurficial portions of presently-existing continental shields. This possibility is considered elsewhere (Tatsch, 1972a: chapter 7).

The post-primordial Earth, in the case of the octantal-fragment mode, evolves essentially from a combination of two processes: (1) the accretion of 5 octants of Earth Prime onto the basic 5400-km primordial Earth; and (2) the differentiation, metamorphism, reworking, igneous activity, and other results of the action of the driving mechanism of the basic model during the past 4.6 billion years (b. y.).

It may be recalled that the driving mechanism of the basic model consists (and has consisted since primordial times) essentially of the resultant of two factors, each of which has been varying in a predictable manner during the past 4.6 b.y.: (1) the potential energy inherent in the geogenetically-disequilibrated shape of the basic 5400-km primordial Earth; and (2) the selectively-channeled energy from the preferential flow of heat, and possibly of volatiles, outward from the three mutually-orthogonal "active" belts formed by the basic octantal fracture pattern of the 5400-km primordial Earth.

Here, as in subsequent discussions and analyses regarding the evolution of the Earth's tectonosphere during the past 4.6 b. y., it is necessary to recall that, in a homogeneous sphere of 6371-km radius, the inertia of the inner 5400-km part normally exeeeds that of the outer 971-km "shell" by at least 60%. Thus, for most purposes, it may be assumed that the disequilibrating inertia of the inner 5400-km part of the model exceeds (and has always exceeded) the equilibrating inertia of the outer 971-km part by a factor of at least 60%. For a non-homogeneous, differentiated body such as the actual Earth, this inertial factor would, of course, normally be even greater than 60%. The significance of this figure is more obvious when some of the mechanics and geometry of the model are discussed in connection with some of the observed behavioral patterns of the actual Earth, such as has been done by the author (Tatsch, 1972a: chapter 8, for example).

The Multiple-Fragment Accretionary Mode.

The multiple-fragment accretionary mode differs from the octantal-fragment mode primarily in the degree to which the 5 "terrestrial" octants of Earth Prime were fragmentized prior to accretion onto the basic 5400-km primordial Earth. For example, if the degree of pre-accretion fragmentation was fairly complete, then it is likely that none of the actual surficial rocks of the present continental shields would now be identifiable with any unmodified portions of the original octants of Earth Prime. This is discussed in greater detail elsewhere (Tatsch, 1972a: chapters 7 and 10).

The Composite-Fragment Accretionary Mode.

The composite-fragment accretionary mode may be considered as an intermediate mode lying between the octantal-fragment mode and the multiple-fragment mode. Depending upon the specific degree of fragmentation assumed for the octants of Earth Prime prior to accretion onto the basic 5400-km primordial Earth, this mode occupies any one of a myriad of intermediate positions within the entire continuum, or envelope, bounded by the octantal-fragment mode and the multiple-fragment mode as extremes.

By extending the limits of the mean, the composite-fragment mode may be made to include one or both of the extremes, thereby collapsing the three modes into two or one, respectively, In this analysis, the three modes will be treated as separate and distinct modes. In the last analysis, any salient differences in the Earth models evolved by the three modes are virtual rather than real; but, by handling the three modes separately, the analysis is more complete and perhaps easier to follow. For this reason, subsequent discussions will assume that there are actually three modes rather than a single mode spread across an entire spectrum, or envelope, including the end points, of the continuum defined by the two extreme modes. The efficacy of this analytical approach is perhaps more obvious when the present continental structures are analyzed in terms of the most likely size of the post-primordial subcratonic blocks that existed within the Earth's tectonopshere, 4.6 b. y. and 3.6 b. y. ago, as well as during the other critical times in the orogenic-cratonic evolution of the Earth's tectonosphere (See, e. g., Tatsch, 1972a: chapters 7 and 15).

The Post-Primordial Earth.

The post-primordial Earth, as derived from the dual primeval planet hypothesis, consists essentially of two parts: (1) an actively-equilibrating, or mobile, 5400-km primordial Earth, differentiated, and fractured basically into octants by three mutually-orthogonal central planes; and (2) an overlying tectonosphere, consisting essentially of the remnants of 5 fragmented octants of Earth Prime accreted onto the 5400-km primordial Earth to a depth of about 1000 km, as modified continually by the driving mechanism of the basic model. Since the primordial Earth loses its identity by becoming incorporated into the post-primordial Earth, it will henceforth be referred to as the Earth's subtectonosphere, or the subtectonospheric portion of the model.

Fig. 2-6 is a schematic sketch of the post-primordial Earth, showing the earliest effects of both igneous and accretionary activity within the tectonosphere. Accretionary activity is indicated schematically at A; igneous activity at I. The octantal blocks of the subtectonosphere are marked S.

Salient Implications of the Basic Tectonospheric Earth Model.

At this point, it is well to list some of the salient implications of the basic Tectonospheric Earth Model in order that the reader may obtain an overall perspective of their nature and scope:

a. Since the 1000-km accretionary material was added to the primordial Earth over a considerable period of time, it follows that various accreted fragments, either individually or in global and regional "shells", would have retained considerable freedom of motion, both with respect to the other accreted fragments and with respect to the basic 5400-km subtectonosphere. Also, it follows that such freedom of motion would most likely be detectable along various planes, surfaces, and shells of preferential rheidity. This is discussed elsewhere (Tatsch, 1972a); there some of these effects are identified as "wave guides", "low-velocity zones", and similar features and phenomena within the tectonosphere. In this connection, it is well to recall that, at a rate of 1 m/yr, a depth of 1000 km is accreted in 1 m. y.; at a rate of 1 cm/yr, it is accreted in 100 million years (m. y.).

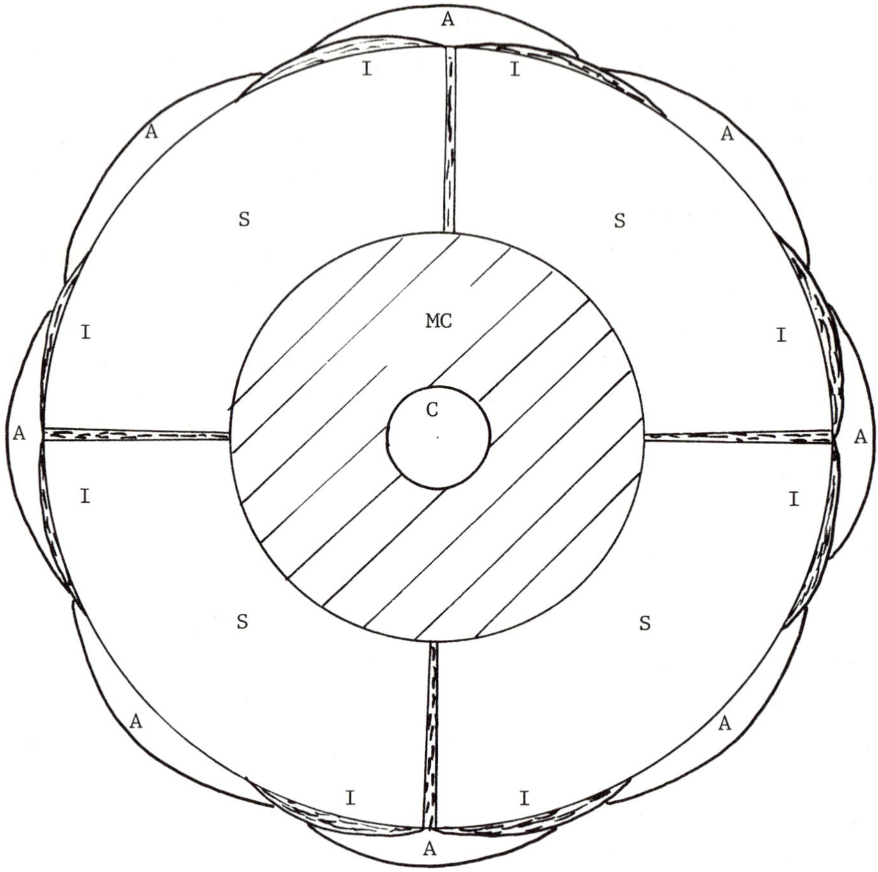

Figure 2-6. The post-primordial Earth, showing the combined effects of primeval igneous and accretionary activities. Accretionary activity is indicated by A; igneous activity by I. The octantal blocks of the subtectonosphere are marked S. MC is the molten core. C is (probably) a small central solid core. Schematic; not to scale. See text for details.

b. The Earth's fracture system consists essentially of the projections, through the tectonosphere, of the basic primordial fracture system of the subtectonosphere. Since the tectonosphere is not homogeneous, such projections are not normally simple, radial projections but somewhat more complex. Other things being equal, the probable error in locating the surficial position of an "energy source" projected from a point on the subtectonosphere through a vertical thickness of 1000 km of average tectonospheric material is roughly 1000 x pi/4, or about 785 km. The significance of this "probable error" figure, as well as the causes for, and magnitudes of, variations in its values are discussed in another work (Tatsch, 1972a; chapter 6), where "cones of activity" are considered in connection with the mechanics and geometry of the evolution of the Earth's tectonosphere during the 4.6 b. y. that the Earth is believed to have been in existence.

c. The Earth's basic deep-fracture system is continuous and worldwide in both the inner (subtectonospheric) and outer (tectonospheric) parts of the model. The latter is basically a consequence of the former, within the constraints of the driving mechanism and the octantally-fractured geometry and mechanics inherent in the model. From this, it appears that the Earth's deep trenches should form a global pattern identifiable with the three mutually-orthogonal great-circular belts of the subtectonosphere. This is discussed elsewhere (Tatsch, 1972a: chapters 6, 7, 8, and 18).

d. "Active" regions (new islands, new mountains, seismic ridges, new volcanoes, etc.) lie roughly above the basic fracture planes of the inner, subtectonospheric part of the model. Due to the effects of planes, surfaces, zones, and shells of preferential rheidity and heat flow, these "active" regions would not, of course, be expected to lie directly above the subtectonospheric fracture system (Tatsch, 1972a: chapter 9).

e. "Inactive" or "fossil" regions (older mountains, older islands, older volcanic belts, seamounts, aseismic ridges, older mineralization belts, etc.) are identifiable with former positions of projections from the subtectonospheric fracture system. To an external observer, these fossil, or inactive, regions appear to have "drifted", in time, with respect to the presently active regions. Normally, such "drifting" involves both translation and rotation, where neither is necessarily linear with respect to time. Also, it might be more accurate to say that the subtectonospheric fracture system, rather than the surficial manifestations thereof, have "drifted", or perhaps even more accurately that both the surficial manifestations and the basic subtectonospheric fracture system have drifted, depending upon whether one wishes to assume a fixed or mobile datum during the analysis. In either case, of course, the results are the same. Also, according to the basic model, the "fossil" features are spatially subparallel to, and temporally derivable from, the "active", or present, features by backward extrapolation procedures. Needless to say, such spatial and temporal relationships are not necessarily linear with respect to time. This is discussed in greater detail elsewhere (Tatsch, 1972a: chapters 7, 8, and 9).

f. Other things being equal, the age of any particular surficial manifestation of the Earth's internal behavior (such as a mountain range, a crustal rift-ridge system, a volcanic belt, etc.) may be approximated roughly by measuring the <u>gross</u> distance by which the particular surficial manifestation has "drifted" from the present position of its genetic subtectonospheric fracture plane. Thus, it is not too surprising that the Appalachian-Caledonian (Upper Paleozoic) orogeny is geographically removed about 30° from the surficial trace of the nearest sub-tectonospheric fracture plane (i. e., Galapagos-Gibraltar), because a drift of 1 cm/yr is roughly equivalent to about 30° per 300 m. y. This is discussed in greater detail elsewhere (Tatsch, 1972a: chapters 7, 8, 9, 10, 12, 13, 16, 17, 18, and 20).

g. The oldest geological features (or surficial manifestations of the Earth's internal behavior), such as Precambrian shields, presently occupy positions identifiable with the basic fracture system of the subtectonosphere. Specifically, they are either: (1) identifiable with the centers of the surficial octants of the subtectonosphere, if stable; or (2) identifiable with the edges of the surficial octants of the subtectonosphere, if they are unstable, active, or in the process of rifting. An example of the latter might be the South American-African "shield" 200 m. y. ago, which has since been rifted and separated as shown schematically in Fig. 2 -7. A more-detailed discussion of this is contained elsewhere (Tatsch, 1972a: chapter 17).

h. Likewise, the newest surficial features of the Earth's internal behavior would be expected to be located near the surficial traces of the three mutually-orthogonal subtectonospheric fracture planes projected to the surface of the Earth.

i. One of the several "unbalanced" conditions existing in the post-primordial Earth resulted from the fact that portions of the Earth's surface must, at various times during the history of the post-primordial Earth, have become squarely juxtaposed above the surficial manifestations of the tectonospheric fracture system. Among the consequences of such a situation would be

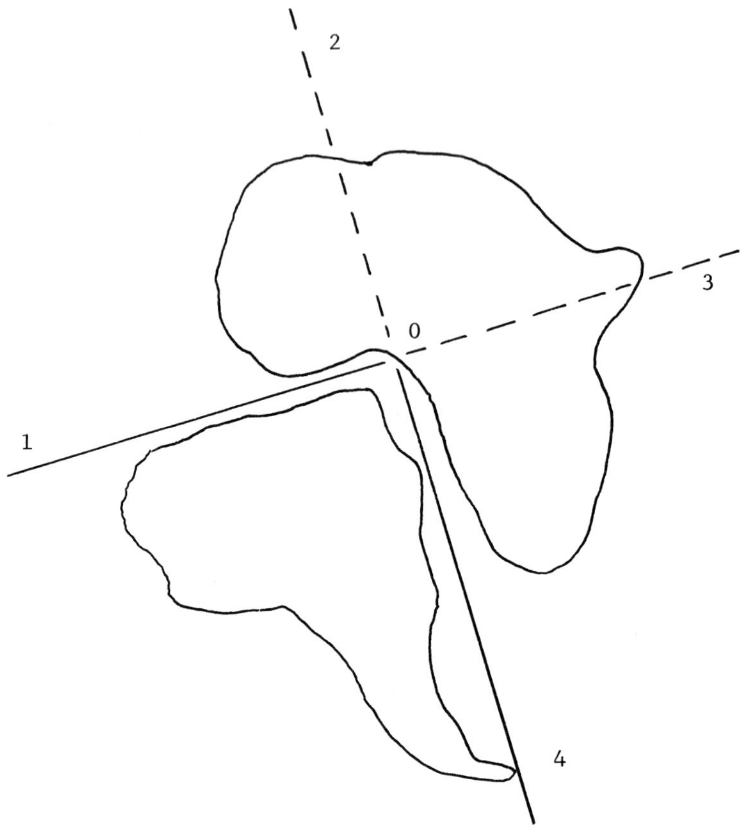

Figure 2-7. Schematic sketch showing two seismotectonomagmatic wedge-belts of activity of the Tectonospheric Earth Model severing Africa from South America during the Mesozoic. The belt segments 1-0 and 4-0 were "tensile" during that part of the Mesozoic; the segments 2-0 and 3-0 were compressive or neutral. See text for details. (Adapted from Tatsch, 1972a).

sea-floor spreading and continental drift (Tatsch, 1972a: chapters 16 and 17).

j. The stress release and other equilibrating effects from the potential energy inherent in the geogenetically-disequilibrated shape of the subtectonosphere are the basic cause of the Earth's deep seismicity (Tatsch, 1972a: chapter 8).

k. Mantle convection may be motivated by the preferential flow of heat along the patterns provided by the subtectonospheric fracture system. Under this concept of the model, mantle convection is a manifestation of the driving mechanism of the basic model. Thus, unlike most other hypotheses, the Tectonospheric Earth Model does not invoke mantle convection as a motive force for sea-floor spreading and continental drift. Rather, mantle convection is derivable from the basic model as an inherent manifestation of the same driving mechanism that drives sea-floor spreading, continental drift, and other omniductive phenomena on the surface of the Earth. The mechanics and geometry of these phenomena, as viewed by the model, are discussed in another work (Tatsch, 1972a: chapters 16 and 18).

l. Igneous activity, both intrusive and extrusive, is motivated by the preferential flow of heat, and possibly of volatiles, outward from the subtectonospheric fracture system (Tatsch, 1972a: chapter 10; 1973a: chapter 4).

m. The preferential flow of heat outward from the subtectonospheric fracture system may be employed to furnish the basic motive power for, and a stable framework for maintaining, a geomagnetic dynamo with a long time constant (Tatsch, 1972a: chapter 14).

n. The roughly antipodal locations of the oceans with respect to the continents is basically a consequence of two factors: (1) the radial translatory motion of the subtectonospheric octantal blocks with respect to the center of the model; and (2) hydrostatic adjustment, or isostasy, on a global scale over a long period of time (Tatsch, 1972a: chapter 15).

o. If viewed on a global scale, certain features such as fault lines, rift-ridges, mountain belts, seamount chains, etc., should normally tend to display "parallel" surficial patterns, intersected by transverse features. As a result, other things being equal, they should tend to form surficial manifestations suggestive of pinnate, grid, and arcuate patterns. These reflect both translation and rotation between the subtectonospheric blocks and the surface of the Earth, most probably along planes, shells, surfaces, and zones of preferential rheidity within the tectonosphere (Tatsch, 1972a: chapters 6 and 18).

The above are a few of the "first order" implications of the basic Tectonospheric Earth Model. The details of these and other implications of the model are discussed elsewhere (Tatsch, 1972a). Higher order implications of the model are the subject of other publications (See, e. g., Tatsch, 1973a through 1973j; 1974b; 1975a; 1975b; 1975c).

<u>Global Surficial Manifestations of the Earth's Internal Behavior Predictable from the Basic Tectonospheric Earth Model</u>.

One of the postulates of the basic Tectonospheric Earth Model is that almost all surficial features and phenomena of the Earth, both past and present, are fundamental consequences of a single, deep-seated driving, or causal, mechanism that has resided within the Earth since the formation of the primordial

Earth, about 4.6 b. y. ago. The details of exactly how such surficial manifestations evolved as a consequence of the behavior of the Earth's tectonosphere during the past 4.6 b. y. is beyond the scope of this book, but many of these details have been discussed elsewhere (See, e. g., Tatsch, 1972a). It may be well to list some of these surficial manifestations of the Earth's internal behavior, in order that the reader may have an appreciation for the general nature and scope of those surficial manifestations that, according to the model, are internally-induced in a predictable manner. These include: global and regional tectonism and orogenesis; sea-floor spreading and continental rifting; global and regional heat-flow patterns; the Earth's magnetic field and its variations, including cyclical but aperiodic reversals in polarity; the Earth's gravity field; volcanic activity; mountain building and related geosynclinal activity; the development of continental shields, mobile belts, fracture systems, rifts, and ridges; continental splitting, separation, and drifting; earthquakes and global and regional seismicity patterns; gravity anomalies and global and regional geoidal patterns; heat-flow anomalies and global and regional thermal patterns; electrical conductivity anomalies; global seismotectonomagmatic and mineralization belts; the Earth's plate-tectonic and omniductive behavior; etc.

At this point it is well to examine one of the surficial manifestations of the Earth's internal behavior, in an attempt to provide a tentative answer to the compound question: Is there any observational evidence that the Earth might, in fact, be representable by the Tectonospheric Earth Model; and, if so, how might the present positions of the eight hypothetical subtectonospheric blocks be identified or located?

To answer this compound question, consideration may be given to a preliminary analysis of any one of several sets of applicable observational data regarding the gross surficial characteristics of the Earth. For example, one such set of data is that showing the locations of the active volcanoes. Since volcanoes appear to be thermally induced and since they are surficial manifestations of tectonospheric activity of some type, they might be expected, according to the model, to lie along segments of global belts defined roughly by the surficial traces of the projections from the three mutually-orthogonal great-circular belts forming the subtectonospheric fracture system.

Therefore, as a first approximation to the location of the surficial manifestations of the subtectonospheric fracture system, we may perform a three-dimensional least-squares analysis of the distribution of the Earth's active volcanoes. Allowing for the probable error expected from the 1000-km projection through the heterogeneous tectonospheric material, we should expect almost all of the Earth's active volcanoes to lie within a surficial distance of about 1000 x $\pi/4 \doteq 785$ km from segments of the radial projections of the subtectonospheric fracture pattern.

When the suggested theee-dimensional least-squares analysis is done, it is found (Tatsch, 1964a) that approximately 93% of the active volcanoes fall within the belts predicted by the model. Specifically, the active volcanoes appear to form segments of three great-circular belts intersecting at the following 6 points, which may be designated tentatively as the "basic tectonospheric points" of the present surficial manifestitions of the basic Tectonospheric Earth Model: 55°N, 165°W (Aleutians); 55°S, 15°E (Bouvet); 5°S, 85°W (Galapagos); 30°N, 5°E (Gibraltar); 5°N, 95°E (Bengal); and 30°S, 175°W (Kermadecs).

Because the probable error of the projection from the subtectonosphere to the surface of the Earth is about $785/111 \doteq 7^{o}$, no purpose is served in attempting to locate the coordinates of the tectonospheric points more "accurately" than about 5° in either latitude or longitude. The geographic names (Aleutians, Bouvet, etc.) that are shown above in parentheses are intended for associative identification of the general areas of the points, rather than as exact geographical descriptions of the points. The associative names will normally be used in subsequent discussions as a matter of convenience; but it should be understood that these points are projections of 6 other points that form a pattern on the surface of the subtectonosphere, 1000 km beneath the surface of the Earth.

For the sake of simplicity, the first two points (Aleutians and Bouvet) may be referred to as the "tectonospheric poles" of the present surficial manifestations of the basic Tectonospheric Earth Model; the other four points (Galapagos, Gibraltar, Bengal, and Kermadecs) may be considered as defining roughly the "tectonospheric equator" of the present surficial manifestations of the basic Tectonospheric Earth Model. Fig. 2-8 shows the above tentative "tectonospheric coordinate system" of the present surficial manifestations of the basic Tectonospheric Earth Model superimposed upon a Mollweide projection of the salient geographical features of the actual Earth. In the figure, the three mutually-orthogonal great circles (ACDFA, ABDEA, and CBFEC) are the radial projections of three similar great circles defining the subtectonospheric fracture system. Each of the intersections at A, B, C, D, E, and F is an orthogonal intersection, and each of the 8 triangles (CBA, CBD, CED, CAE, FEA, FDE, FBD, and FBA) is an equilateral spherical triangle (i. e., with all sides and angles equal to 90°). On the surface of the tectonosphere, 90° equals to about 10,000 km. On the surface of the subtectonosphere, 90° equals about 8500 km.

The great circle, ABDEA, also marked QQQQ in Fig. 2-8, is a projection of the subtectonospheric equator and it defines roughly the tectonospheric equator of the present surficial manifestations

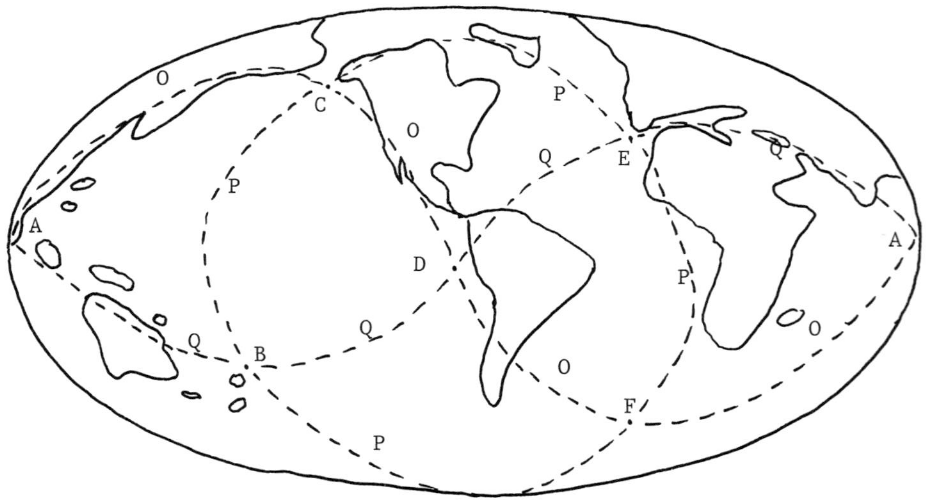

Figure 2-8. Tentative "tectonospheric coordinate system" of the present surficial manifestations of the basic Tectonospheric Earth Model superimposed upon a Mollweide projection of the salient geographical features of the actual Earth. See text for explanation of symbols.

of the basic Tectonospheric Earth Model. The great circle, CBFEC, also marked PPPP, is a projection of the subtectonospheric prime meridian and it defines roughly the tectonospheric prime meridian of the present surficial manifestations of the basic Tectonospheric Earth Model. Similarly, the great circle, ACDFA, also marked OOOO, is a projection of the subtectonospheric orthogonal meridian.

A tectonospheric coordinate system based on the above is particularly useful as a base for performing harmonic analyses of the Earth's surficial manifestations. According to the dual primeval planet hypothesis, the Earth's surficial manifestations are tectonospherically induced and, therefore, "harmonic" with respect to the tectonospheric coordinate system, rather than with respect to the geographic coordinate system normally used for harmonic analyses when using other models.

The Tectonospheric Earth Model as a Supplementary Tool in the Exploration for Minerals.

Briefly stated, the Tectonospheric-Earth-Model approach to mineral exploration is based upon a straightforward premise: if one knew today where all the Earth's minerals were 4.6 b. y. ago, and if he could define the geometrical, mechanical, thermal, and chemical behavior of the Earth during the past 4.6 b. y., then he should be able to predict the present locations of the Earth's mineral deposits (Tatsch, 1973a).

The following chapters discuss how the Tectonospheric Earth Model may be used to define the complex geometrical, mechanical, thermal, and chemical interrelations that exist among igneous activity, folding, faulting, uplifting, subsidence, and other omniductive geophenomena that have contributed to the origin, evolution, and present characteristics of gold deposits.

Chapter 3

THE GEOMETRICAL, MECHANICAL, THERMAL, AND CHEMICAL BEHAVIOR OF THE EARTH DURING THE PAST 4.6 BILLION YEARS

The Tectonospheric Earth Model differs from simple "plate-tectonics" concepts in two fundamental respects: (1) it extends the applicable spatio-temporal domain to a depth of at least 1000 km and to a span of about 4.6 b. y.; and (2) it provides a single, long-lived, deep-seated, global driving mechanism that has been motivating the Earth's internal behavior during the 4.6 b. y. that the Earth is believed to have been in existence (Chapter 2; Tatsch, 1972a). It is well to consider, at this point, some of the aspects of these two extensions that have been provided by this model.

For ease of presentation, three aspects of the Earth's surficial features and phenomena may be considered: (1) a global view of the Earth's present surficial features; (2) the most probable surficial features of the proto-Earth 4.6 b. y. ago; and (3) the evolution of the Earth's present surficial features from those of the proto-Earth during the past 4.6 b. y. Because the Earth's evolutionary period appears to have been interrupted about 3.6 b. y. ago, the total time period will be considered in 2 parts: (1) 4.6 b. y. ago to 3.6 b. y. ago; and (2) 3.6 b. y. ago to the present.

A Global View of the Earth's Present Surficial Features and Phenomena.

One of the theses of the Tectonospheric Earth Model is that the Earth's surficial features and phenomena are a reflection of its internal behavior. It is well to review some of these features and phenomena as they are interpreted in terms of the Tectonospheric Earth Model concept:

a. Orogenic-cratonic structure of the continents. The

orogenic-cratonic structure of all continents appears to be more-or-less the same (Tatsch, 1972a: chapter 7). What is seen, in general, is an old Precambrian shield successively surrounded by long rectilinear orogenic, geosynclinal, or "seismotectonomagmatic" belts, often broken or otherwise modified by overprinting by subsequent seismotectonomagmatic activity, with at least 3 orogenies fairly well preserved in at least some continents; e. g., the Appalachian, Innuitian, and Cordilleran orogens of the Canadian shield. Unanswered are questions as to why the orogenic belts are rectilinear and of different ages within any one continent but appear to to extend worldwide, with their activity having occurred in cyclical-but-aperiodic episodes for at least the past 3.6 b. y., and with concentrations of this activity having occurred at about 0.9 b. y. intervals, or at sub-multiples thereof. Also not understood is why the spatio-temporal relationships within all orogenic belts are remarkably similar, regardless of where found or to which period they belong. However, in spite of these unanswered questions, the orogenic-cratonic structure of the continents suggests a remarkably orderly behavior over an extremely long period of time, perhaps as long as 3.6 b. y. or even 4.6 b. y. This, in turn, suggests that the continental structures are surficial manifestations of a long-lived, deep-seated, global driving mechanism.

b. The Earth's seismicity and earthquakes. When considered as a surficial manifestation of the Earth's internal behavior, earthquakes suggest that the Earth's internal behavior is controlled by a global mechanism completely independent of continental margins, island arcs, and similar features, except insofar as they may also be surficial features of the same mechanism (Tatsch, 1972a: chapter 8).

c. Heat-flow and other geothermal activity. When viewed

on a global scale, it appears that the Earth's internal behavior reveals itself through thermal manifestations on the surface. The exact correlation between internal behavior and surficial manifestations of the thermal Earth is not clearly understood, perhaps because of the paucity of global heat-flow data. If a generalization may be made, it is that regions of modern orogenesis correlate with high heat-flow rates (Tatsch, 1972a: chapter 9).

d. Intrusive and extrusive activity. When viewed on a global scale, intrusive and extrusive activity, of both the mid-ocean-ridge type and the Alpine-ultramafic-belt type, may be considered to be surficial manifestations of a single, global mechanism characterized by cyclical-but-aperiodic activity along long, linear belts of deep-seated control. Closer scrutiny reveals that intrusive and extrusive activity forms a complete spatio-temoral spectrum, or continuum, within each cycle of orogenic activity whether of the mod-ocean-ridge type or of the Alpine-ultramafic-belt type (Tatsch, 1972a: chapter 10).

e. Mountains and geosynclines. The relationship that exists between mountains and geosynclines is not completely understood. It appears, however, that, when geosynclines and mountains are analyzed simultaneously on a global scale, then their origin and evolution appear to be controlled by a single, global mechanism deep within the Earth (Tatsch, 1972a: chapter 12).

f. The Earth's gravity field and the shape of the geoid. On the basis of present evidence, it appears that the Earth's gravity field and the shape of the geoid are surficial manifestations of the internal behavior of the Earth during the past 4.6 b. y. However, it appears also that mass redistributions lag behind the forces causing the Earth's internal behavior. The shape of the geoid should correlate, therefore, with the tectonospheric driving mechanism rather than with purely surficial manifestations of the Earth's internal behavior (Tatsch, 1972a: chapter 13).

g. Geomagnetism and polarity reversals. When the Earth's magnetic field is considered over a long period ot time, several tentative conclusions may be drawn: (1) the geomagnetic dynamo has evolved, and is presently driven, by the same global causal mechanism that has produced the Earth's internal behavior during the past 4.6 b. y.; (2) the geomagnetic field correlates, now and in the past, with global characteristics of the Earth's driving mechanism; and (3) the cyclical-but-aperiodic polarity reversals are definable in terms of the geometry and mechanics of the Earth's driving mechanism (Tatsch, 1972a: chapter 14; 1974c). On this basis, it appears that, when viewed on a global scale, then geomagnetism and polarity reversals are surficial manifestations of the Earth's internal behavior.

h. Continents and oceans. Viewed on a global scale and over a long period of time, it appears that oceans and continents are surficial manifestations of the Earth's internal behavior (Tatsch, (Tatsch, 1972a: chapter 15).

i. Crustal rifting and sea-floor spreading. Crustal rifting (and sea-floor spreading as a subsidiary phenomenon thereof) appears to be a surficial manifestation of the Earth's internal behavior. Furthermore, such rifting (and spreading) appears to occur when and wherever the Earth's internal behavior causes crustal tension (Tatsch, 1972a: chapter 16).

j. Plate tectonics and related matters. If plate motions are surficial manifestations of the Earth's internal behavior, then the origin, evolution, and present characteristics of the plates should be correlatable with certain features and phenomena of the Earth's internal behavior (Tatsch, 1972a: chapter 18).

k. Continental drift and polar wandering. When viewed on a global scale over a long period of time, it appears that continental drift and polar wandering are surficial manifestations of the Earth's internal behavior, during possibly the entire existence of the Earth, but that the geometry and mechanics of the Earth's interior and the nature of the driving mechanism are difficult to define in terms of simple plate-tectonics concepts (Tatsch, 1972a: chapter 17).

In analyzing the present surficial features of the Earth that might have been caused by possible internal-behavior patterns of the Earth during the past 4.6 b. y., it is well to consider briefly the possible surficial configuration of the Earth 4.6 b. y. ago and at various critical times since then.

The Earth's Most Probable Surficial Features 4.6 b. y. Ago.

The configuration of the primordial Earth's most probable surficial features depends upon exactly which hypotheses are adopted to analyze the evolution of the Earth from the solar nebula (Tatsch, 1972a: chapter 1) and to explain the evolution of the Earth-Moon system (Tatsch, 1974a). If it is assumed that the Earth accreted from the solar nebula and that the Moon formed in the vicinity of (but not part of) the Earth, then the Earth's surface would probably have been devoid of any significant topographic features other than those that would be genetic to an accretionary process. Thus, the Earth's surface would have presented a more-or-less homogeneous aspect 4.6 b. y. ago, both from the standpoint of topography (i. e., smooth) and from the standpoint of composition (i. e., more-or-less the same) over the entire surface.

If, on the other hand, it is assumed that the Earth's surficial composition and/or topography were not homogeneous 4.6 b. y. ago, then such features would most probably not have survived to this date without some internal of external rejuvenation at least on one occasion since then. For the purpose of this analysis, it may be assumed that the Earth's surface 4.6 b. y. ago was devoid of any features other than those that would have been genetic to an accretionary formation from the solar nebula at that time (Tatsch, 1974a).

The Earth's Most Probable Surficial Features 3.6 b. y. Ago.

If it is assumed that the Earth had no appreciable surficial features 4.6 b. y. ago, when were the surficial features first acquired, and what caused them to form? Were they acquired through external or internal behavior? If external, what was it? If internal, why did it occur when it did, rather than sometime eralier or later?

If the Earth's surficial features resulted from external sources or causes, during the period 4.6 b. y. to 3.6 b. y. ago, then such features would not now be active but would resemble the Archean shields and would have an age of about 3.6 b. y.

If the Earth acquired its surficial features from internal sources or causes, during the period 4.6 b. y. to 3.6 b. y. ago, then such features would not now be active unless the same causative internal hehavior had been operating ever since 3.6 b. y. ago. In that case, it would be reasonable to assume that the internal mechanism has been operating more-or-leas continuously during the past 4.6 b. y. (because nothing internal to the Earth would have been available to activate it 3.6 b. y. ago).

In any case, it appears that the Precambrian shields existed 3.6 b. y. ago, but not necessarily in their present forms and locations. Also, it appears that these shields have existed in one or more locations and forms ever since then, without having been severely heated or shocked except at their peripheries.

The Evolution of the Earth's Present Surficial Features.

Regardless of whether it is assumed that the Earth's surficial features of 3.6 b. y. ago were externally or internally induced, we must consider certain questions about what has happened to the Earth's surface since then. For example, have there been significant externally-induced catastrophes that migth have altered the Earth's surficial features during the past 3.6 b. y.? If so, what were they, when did they occur, and what caused them? If no such external influences have operated during the past 3.6 b. y., is it reasonable to assume that all modifications to the Earth's surface since then were induced by the Earth's internal behavior?

To answer these and related questions about the evolution of the Earth's present surficial features, global-scale analyses must be made regarding both the present and past internal behavior of the Earth. Several international scientific bodies have been devoted to such analyses, e. g., the International Upper Mantle Commission and the Interunion Commission on Geodynamics (Drake, 1970, 1973). Other studies have been made on a national or institutional basis (See, e. g., Anderson, 1975; Bullen, 1974a, 1974b; Irving and McGlynn, 1975; Kaula, 1975; Berg and Sutton, 1975; Oyarzun, 1975; Cabre, 1975; Schilling and Bonatti, 1975; Yoshii, 1975; Vogt, 1975; Richter and Parsons, 1975; Froidevaux and Schubert, 1975; Rosendahl et al., 1975; Oldenburg and Brune, 1975. See also the many earlier references that are cited in Tatsch, 1975a: pages 78 and 79).

When the Earth is considered as a long-lived, global entity, the results of the above, and other, analyses indicate that most past and present geophenomena are manifestations of the Earth's internal behavior during the past 4.6 b. y., but that the geometry and mechanics of the causative internal behavior and the nature of the ultimate driving mechanism remains to be defined. It is the writer's hypothesis that the Tectonospheric Earth Model, derived from a dual primeval planet hypothesis (Tatsch, 1972a) provides some answers for the undefined geometry and mechanics and for the nature of the Earth's internal driving mechanism.

Chapter 4
PLATE TECTONICS, OMNIDUCTIVE PROCESSES, AND SEISMOTECTONOMAGMATIC BELTS

It has been known for some time that gold deposits are often found near plate boundaries and near other surficial vestiges of the Earth's internal behavior. One of the purposes of this chapter is to attempt to generalize this observation into a long-lived, deep-seated concept that may best be expressed in terms of "omniductive processes" and "seismotectonomagmatic belts". This concept forms one of the bases for using the Tectonospheric Earth Model as a supplementary tool in the exploration for gold deposits.

Plate Tectonics, Gold Deposits, and the Tectonospheric Earth Model.

Before considering the applicability of the Tectonospheric Earth Model as a supplementary tool in the exploration for gold deposits, it is well to condider some of the implications of the simple plate-tectonics concept and to review briefly how the Tectonospheric Earth Model interprets this concept and its implications in the exploration for gold deposits.

Specific analyses regarding the simple plate-tectonics concept have been summarized in the literature (See Tatsch, 1975a, page 80, for a listing of the more recent summaries of such). Suffice it here to say merely that the simple plate-tectonics concept leaves several unanswered questions, both from the standpoint of a general tectonic hypothesis and in its applicability to gold and other mineral deposits. These unanswered questions have been adequately discussed elsewhere (See, e. g., Tatsch, 1975a, pages 81 through 88, which cite and discuss all significant references through early 1975).

The Simple Plate-Tectonics Concept as a Supplementary Tool in the

Exploration for Gold Deposits.

Many characteristics of the Earth's gold deposits have defied attempts to define them in terms of a single michanism that has been operating within the Earth during the 4.6 b. y. that the Earth is believed to have been in existence. Several years ago, the simple plate-tectonics concept received some attention in this connection. But it is now generally agreed by explorationists that the simple plate-tectonics concept, per se, does not provide the final answer.

There are several resasons for this deficiency of the simple plate-tectonics concept. Many recent observations that have a bearing on the geometrical, mechanical, thermal, and chemical aspects of the Earth's behavior appear to be at variance with the simple plate-tectonics concept. Some of these have been discussed recently in a related context (Tatsch, 1975a: pages 82 through 88, which include all significant references through 1974).

The Migration of the Patterns of the Surficial Manifestations of the Earth's Internal Behavior.

Very little is known about the details of the geometrical, mechanical, thermal, and chemical aspects of the Earth's internal behavior. However, many observers feel that the details of the Earth's internal behavior are best analyzed in terms of the panoramic changes that have occurred in the geophenomena and features at the Earth's surface. Even more is revealsd when these surficial patterns are observed to migarate within a spatio-temporal frame that extends through a span of billions of years.

It is well to summarize, therefore, what is known about the migration of patterns of the surficial manifestations of the Earth's internal behavior:

a. Bowin (1974) has reported a pattern of plate motion

that have migrated from the South Pacific into the Indian Ocean during the Cretaceous and early Tertiary, followed by a migration across the Indian Ocean during the Tertiary.

b. Some observers feel that the opening of the Gulf of Aden may be a more-recent and extended surficial manifestation of the migration reported by Bowin (1974).

c. Based on observational evidence, these migrations comprise the episodic formation of new lineaments along which sea-floor spreading and other omniductive phenomena occur (See, e. g., Sclater and Fisher, 1974).

d. In many cases, these new lineaments of seismotectono-magmatic activity are both translated and rotated from erstwhile spreading lineaments or axes (Tatsch, 1972a: page 739; Bowin, 1974).

e. In most of these, the development of the new lineaments or axes seems to follow deep-seated, long-lived patterns that might have existed deep within the Earth for billions of years.

f. These and other observations from all parts of the Earth suggest that the surficial manifestations of the Earth's internal behavior extend far beyond the current (or even the recent-past) boundaries of the Earth's surficial plates (See, e. g., Bowin, 1974).

A closer scrutiny of the above suggests that many questions remain unanswered. Why, for example, does the simple plate-tectonics concept, which appears fairly versatile when applied at the global and regional levels, fail at the local level? Does this mean that the observations are inadequate? Or does it mean that the simple plate-tectonics concept is not the final answer?

Global Seismotectonomagmatic Belts.

A close scrutiny of the Earth's surface reveals that certain elongate belts, characterized by varying degrees of seismic, tectonic, and magmatic activity, have existed in various regions of the Earth during most of the 4.6 b. y. that the Earth is believed to have been in existence. These belts may be referred to as seismotectonomagmatic (STM) belts.

Much evidence suggests that all these seismotectonomagmatic belts are similar, regardless of when and where formed (See, e. g., Marlow et al., 1973, who have considered island arcs, orogenic belts, and certain other tectonomagmatic episodes that fall into the category of this definition). Other recent studies of the various facets of the geometrical, mechanical, thermal, and chemical aspects of some of the seismotectonomagmatic belts, of various ages and in various parts of the world, have been completed. These have been summarized elsewhere (Tatsch, 1975a: pages 90 through 115). They need not be repeated here. Suffice it merely to list some of the Earth's features and phenomena that have been interpreted in terms of the seismotectonomagmatic-belt concept of the Tectonospheric Earth Model, where the following listed listed page numbers refer to Tatsch, 1975a:

a. The Earth's earliest Precambrian activity: page 90.

b. The evolution of Australia: page 94.

c. The development of the Earth's earliest orogenic belts: page 95.

d. The development of the Earth's sialic crust: page 97.

e. The evolution of mountain belts: page 100.

f. The development of faunal assemblages: page 102.

g. Continent-continent collisions: page 102.

h. Descending lithospheric plates: page 104.

i. Transform faults: page 104.

j. Inter-arc basins: page 106.

k. Omniductive processes and ultramafic rocks in paired belts: page 108.

l. Gravity tectonics: page 108.

m. The observed seismicity of the Baffin Bay region: page 110.

n. Alpine peridotite types: page 110.

o. The geologic development of the Indonesian arcs: page 111.

These and other examples show how the seismotectonomagmatic-belt concept may be used when the simple plate-tectonics concept proves to be inadequate, weak, non-committal, or not applicable when analyzing the Earth's observed features and phenomena.

Mineral Deposits as Interpreted by the Seismotectonomagmatic-Belt Concept of the Tectonospheric Earth Model.

As indicated earlier, many mineralizations appear to be aligned in belts that are closely associated with plate-tectonic boundaries and with certain other tectonomagmatic, seismotectonic, and omni-ductive features of the Earth's surface. The exact nature and cause for this association between mineralizations belts and the geological structure and behavior of the Earth are not known. One of the purposes of this book is to make a preliminary analysis of this association between gold mineralization belts and the internal behavior of the Earth during the past 4.6 b. y.

One of the premises of the book is that the features and phenomena of the Earth's surface reflect the geometrical, mechanical, thermal, and chemical aspects of the Earth's internal behavior during the 4.6 b. y. that the Earth is believed to have been in existence. By studying the behavior of the Earth's surface, one can determine, hopefully, some of the aspects of the Earth's internal behavior.

One of these features and phenomena of the Earth's surface that may reflect the geometrical, mechanical, thermal, and chemical aspects of the Earth's internal behavior is the spatio-temporal panorama presented by the origin, evolution, and accumulation of mineral deposits during the past 4.6 b. y. (Tatsch, 1973a; 1974b; 1975a). One aspect of this panorama reveals itself in the relationship of the local, regional, and global patterns of metallogenesis to the corresponding patterns of the development of the seismo-tectonomagmatic belts.

Closer scrutiny of the Earth's surface suggests that local, regional, and global metallogenic zoning is produced by (1) different structural environments, (2) crust and mantle heterogeneities, and (3) different histories of evolution (See, e. g., Radkevich, 1972). Most local zoning relfects the influence of episodic intruso-extrusive activity plus the geometrical, mechanical, thermal, and chemical aspects of the surficial manifestations of the Earth's internal behavior. Many regional metallogenic zones are constituted similarly in all parts of the world; that is, (1) gold and copper in eugeosynclines; (2) tin and tungsten in terrigenous miogeosynclines; and (3) lead and zinc in geanticlines comprising limestone.

One example of global zoning may be seen in the circum-Pacific belt, which is separable into an inner and an outer megazone. The inner megazone, with Neogene andesitic magmatic rocks and mineralization of the femic type (Cr, Ni, Cu, Au), originated close to the ocean on crust of oceanic and transitional type. The outer megazone was developed on crust of continental type, being characterized by mineralization of sialic affiliation (Sn, W) related to the Mesozoic granites and mineralization of femic affiliation in eugeosycliess (North and South American cordillera).

These and other generalizations, when integrated into the seismotectonomagmatic-belt concept and combined with observations regarding the Earth's internal and external activity, serve as the basis of interpreting the origin, evolution, and characteristics of gold deposits during the 4.6 b. y. that the Earth is believed to have been in existence. These interpretations are made in Chapters 6 through 10. Before doing that, however, it is well to take a brief look at the concept of using observations regarding intruso-extrusive activity as a long-lived, deep-seated basis for the origin, and evolution of mineral deposits. This is done in the next chapter.

Chapter 5

INTRUSIVE AND EXTRUSIVE ACTIVITY: A LONG-LIVED, DEEP-SEATED BASIS FOR THE ORIGIN AND EVOLUTION OF MINERAL DEPOSITS

Many intrusive and extrusive episodes appear to be comagmatic even when not contemporaneous. In some of these episodes, ultramafic rocks have been wholly or partly intruded after the main metamorphic and deformational phases of the tectonomagmatic episode. In other episodes, this sequence has been reversed. In still others, intrusive and extrusive phases have been pene-contemporaneous with the metamorphic and deformational phases. The Kenoran orogeny (2.73 to 2.33 b. y. ago) displays several distinct igneous, metamorphic, deformational, and faulting phases within the Canadian Shield (See, e. g., Weber and Stephenson, 1973). In other times and places, similar phases occur but not necessarily in this same sequence.

Intrusive and extrusive activity is closely associated spatially and temporally with tectonic activity. The intrusion of voluminous basaltic dikes in the volcanic and batholithic structures often continues before, during, and after deformation, suggesting thereby a continuous process of basic and acidic magmatism contemporaneous with tectonic activity (See, e. g., Strong, 1973). All alkali olivine basalt series are similar to other alkali olivines regardless of whether formed in deep-oceanic, island-arc, or continental environments (See, e. g., Schwarzer and Rogers, 1973).

These and other recent observations suggest that the various types of intruso-extrusive activity, as well as the associated tectonic activity, are motivated by a long-lived, deep-seated, global mechanism that has been operating within the Earth during most of the 4.6 b. y. that the Earth is believed to have been in existence. It is well to examine these and other aspects of the intrusive and extrusive episodes, in an attempt to define these episodes in terms of their constituting a long-lived, deep-seated basis for the origin and evolution of mineral deposits.

Intrusive and extrusive activity has played an important role in the evolution of the Earth's tectonosphere during the past 4.6 b. y. This intrusive and extrusive activity includes volcanic activity, the creation of basaltic traps, both in continental and in oceanic areas, mantle upwelling at crustal rifts, and the emplacement of batholiths, laccoliths, and other intrusive bodies during the past 4.6 b. y.

This book considers intrusive and extrusive eruptives as a single entity. This is done in order to emphasize the concept that a single global deep-seated, long-lived mechanism is responsible for the behavior of both of these types of eruptives. The term "eruptive" is used in its general sense (Challinor, 1964), i. e., including both intrusive and extrusive activities involved in breaking through from one level to another within the tectonosphere. Thus, an eruptive that "breaks through" to a subsurface level is an intrusive; one that breaks through to the surface, an extrusive. An intrusive during one orogeny may become an extrusive through the action of a subsequent orogeny.

Arthur Holmes pointed out many years ago that the three major problems of igneous petrology are: (1) the nature and source of primary magmas; (2) the mechanisms of magmatic intrusion and emplacement; and (3) the processes that have brought about the manifold diversity of igneous rocks. One additional problem may be added to these: (4) the nature and source of the energy for the ascensive mechanism. These four problems have been discussed elsewhere (Tatsch, 1973a: chapter 4).

Intrusive and Extrusive Activity as Interpreted by the Tectonospheric Earth Model.

Most observers agree that present mineral deposits have been derived from materials that have existed within the Earth for at least 3.6 b. y. But there is no general agreement on the evolutionary processes that have led to the localization of minerals. Many layered deposits of metallic minerals were contemporaneous with the formation of the enclosing rock. Other minerals were introduced later, perhaps through some geometrical, mechanical, thermal, or chemical aspect of those particular rocks that favored a preferential path for the mineralizing materials. Still others were redistributed from earlier mineralizations during subsequent tectonothermomagmatic activity. These redistributions, in some cases, caused enrichment; in others, impoverishment (Bain, 1968).

According to the Tectonospheric Earth Model, the origin, evolution, and present characteristics of mineral deposits are closely associated with the Earth's intrusive and extrusive activity, which, in turn, is associated with orogenic activity. These associations may be seen in the evolution of most continents.

The geological history of Africa, for example, bears a close relationship to the origin, evolution, and present characteristics of mineral deposits within that continent. The history of most continents can be studied best as an orogenic-cratonic evolution. Thus, we find that Africa has undergone at least seven major orogenic events, each of which had a profound effect on the origin, evolution, and present characteristics of the mineral deposits within the African continent: (1) Transvaalian, 3 b. y. ago; (2) Shamvaian, 2.8 to 2.5 b. y. ago; (3) Eburnian-Huabian, 1.85 b. y. ago; (4) Kilbaran, 1.1 b. y. ago; (5) Damaran-Katangan (Pan-African), 0.55 b. y. ago; (6) Acadian-Hercynian, 0.40 to 0.20 b. y. ago; and (7) Alpine, 0.18 to 0.02 b. y. ago.

In Africa, as in other continents, the earliest orogens appear to have been less affected by more recent orogenies. Therefore, the mineral deposits within the oldest orogenic belts can be expected to differ from those within more recent orogenic belts. Thus, the older orogenic belts (long since welded into the present cratons) contain Africa's principal deposits of asbestos, chromium, diamonds, gold, and iron. The younger orogens contain Africa's principal deposits of beryllium, cobalt, copper, lead, niobium, tantalum, tin, and tungsten.

According to the Tectonospheric Earth Model, all minerals that have been deposited during the past 4.6 b. y. have originated within an active seismotectonomagmatic belt (Tatsch, 1972a). Thus, Africa's primary deposits of asbestos, chromium, diamonds, gold, and iron originated within seismotectonomagmatic belts that controlled the three earliest major African orogenies, i. e., the Transvaalian, Shamvaian, and Eburnian-Huabian. Similarly, Africa's primary deposits of beryllium, cobalt, copper, lead, niobium, tantalum, tin, and tungsten became concentrated within the seismotectonomagmatic belts that controlled the four more-recent major African orogenies, i. e., the Kilbaran, Pan-African (Damaran-Katangan), Acadian-Hercynian, and Alpine.

Almost all of the Earth's oldest basic seismotectonomagmatic belts have long since become welded to the cratons (and in most cases are now considered to be part of the respective cratons). However, the Earth's present metallogenic provinces attest to the erstwhile activity of these fossil seismotectonomagmatic belts. In some cases, the process by which these erstwhile active seismotectonomagmatic belts became welded to the respective cratons has caused the original deposits within the older fossil seismotectonomagmatic belts to become less clear-cut than is characteristic within the more-recently active seismotectonomagmatic belts. Thus, minerals characteristic of the older seismotectonomagmatic belts occur where large segments of these older seismotectonomagmatic belts have been involved in younger orogenic events. Consequently, a number of asbestos, gold, and iron deposits found near the margins of younger orogens probably owe their present distribution to this later orogenic activity.

In some cases, minerals originally emplaced in fossil seismotectonomagmatic belts have been transported to deposits within younger seismotectonomagmatic belts. One example concerns the diamondiferous gravels that have been transported by the Orange River into Tertiary and Recent sediments. Similar examples may be found in other continents involving different minerals at different times.

Intrusive and extrusive activity, according to the Tectonospheric Earth Model, is associated with the "cones of activity" as well as with the seismotectonomagmatic belts of the model (Tatsch, 1972a: chapter 6). That is, intrusive and extrusive activity due to a given energy source (existing at the surface of the subtectonosphere) would be expected to occur somewhere within a cone of activity centered on the "energy source" as a vertex and intersecting the surface of the Earth in a circle with a radius of approximately 785 km (= 1000 pi/4) in the idealized case (Tatsch, 1972a: figure 6-2e-2, chapter 6).

When local cones of activity are formed for every point on the subtectonospheric fracture system, then the surficial trace of their vertexes consists of 3 mutually-orthogonal great circles, aggregating about 102,000 km in length. The resulting cones of activity form an infinity of cones whose "envelope" may be described as 3 global tectonospheric seismotectonomagmatic belts. At the surface of the Earth, where the radius is 1.18 times that at the base of the tectonosphere, the centerlines of these cones form 3 mutually-orthogonal great circles aggregating 102,000 x 1.18 = 120,000 km (Tatsch, 1972a: chapter 6).

All intrusive and extrusive activity, according to the model, is motivated by the Earth's driving mechanism. Intrusive and extrusive activity resulting directly from the Earth's driving mechanism would normally be expected to occur within the present position of an active seismotectonomagmatic belt; that resulting indirectly from the Earth's driving mechanism may occur anywhere within the Earth without regard to present positions of the active seismotectonomagmatic belts (Tatsch, 1972a).

Because, at any given time, active or effective energy sources may be expected along about 50% of the subtectonospheric fracture system (i. e., along about 51,000 km), intrusive and extrusive activity would be expected to occur within about half of the global seismotectonomagmatic belt system (i. e., along "belts" aggregating about 60,000 km on the surface of the Earth) at any one time. Intrusive and extrusive activity would not necessarily occur simultaneously in all active segments of the global seismotectonomagmatic belts. Also, some types of magmatic activity normally would not accompany certain types of tectonic activity. For example, ultrabasic effusives would not normally accompany compressive tectonics according to the model. Nor would more than one specific type of magmatism be expected to occur at a given time at a given location (Tatsch, 1972a).

Crustal and Upper-Mantle Evolutionary Modes Postulated by the Tectonospheric Earth Model.

Depending upon what basic assumptions are made, the Earth's crust and upper mantle developed, according to the Tectonospheric Earth Model, through any one of three basic modes of accretionary evolution: (1) octantal-fragment, (2) multiple-fragment, and (3) composite-fragment (Tatsch, 1972a: chapter 3). The type and behavior of magmatic activity during the past 4.6 b. y. would have been different in each of these three accretionary modes of tectonospheric evolution. These may be summarized.

In the octantal-fragment mode for the evolution of the Earth's tectonosphere, the 5 "terrestrial" octants of Earth Prime are not completely fragmentized prior to accretion onto the basic 5400-km primordial Earth (Tatsch, 1972a: chapter 3). In such a case, parts of each of the 5 terrestrial octants of Earth Prime might still exist today as unmodified portions of the Earth's present tectonosphere, perhaps even as subsurficial portions of presently-existing continental shields.

If the Earth's tectonosphere evolved through the octantal-fragment mode, the intrusive and extrusive activity during the past 4.6 b. y. would have been somewhat different from that in the multiple-fragment and composite-fragment modes for the evolution of the Earth's tectonosphere. The details of these differences have been discussed (Tatsch, 1972a), but they need not be considered further here.

The multiple-fragment mode differs from the octantal-fragment mode primarily in the degree to which the 5 "terrestrial" octants of Earth Prime were fragmentized prior to accretion onto the basic 5400-km primordial Earth. If the degree of pre-accretion fragmentation was fairly complete, then it is not very likely that any of the actual surficial rocks of present continental shields would now be identifiable with unmodified portions of the original octants of Earth Prime, because intrusive and extrusive activity during the past 4.6 b. y. would have modified these multiply-fragmentized primordial rocks.

The composite-fragment mode may be considered as an intermediate mode lying between the octantal-fragment mode and the multiple-fragment mode. Depending upon the specific degree of fragmentation assumed to have occurred within the octants of Earth Prime prior to accretion onto the basic 5400-km primordial Earth, this mode occupies one of a myriad of intermediate positions within the entire spectrum, or envelope, bounded by the octantal-fragment and multiple-fragment modes as extremes. A similar statement may be made about the modification of these fragments by intrusive and extrusive activity during the past 4.6 b. y. That is, the amount and degree of modification by intrusive and extrusive activity in the composite-fragment mode would be intermediate between that in the octantal-fragment and that in the multiple-fragment mode of tectonospheric evolution.

Intrusive and Extrusive Activity as a Function of Paths of Preferential Heat Flow within the Tectonosphere.

Within the seismotectonomagmatic belts, the nature and location of intrusive and extrusive activity depends largely upon: (1) the nature of the local stress field and (2) the paths of preferential heat flow (Tatsch, 1972a: chapter 6).

If an energy source at some point along the subtectonospheric fracture system consisted of heat, and if the tectonosphere above that point were homogeneous, then it might be a relatively simple matter to determine the most probable paths that the heat might follow through the tectonosphere to the surface of the Earth. Because the tectonosphere (and specifically the seismotectonomagmatic belt) is not homogeneous, the preferential heat-flow path (from the surface of the subtectonosphere to the surface of the Earth) might be quite circuitous in order to circumvent heterogeneities of low thermal conductivity. The exact path followed by the heat would influence the nature and location of the resulting magmatic activity, whether intrusive or extrusive. This, according to the model, accounts for the circuitous routes taken by some feeders, or conduits, to volcanoes and to intrusive bodies (Tatsch, 1972a: chapter 9).

Paths of preferential heat flow within the Earth's tectonosphere are of interest to the origin, evolution, and present characteristics of gold deposits because most paths of preferential heat flow are closely associated with paths of preferential flow of volatiles and magmas. The importance of these associations may be emphasized by considering them in connection with the origin of mineral deposits.

Thus, we see that magmas appear to be the ultimate source of essentially all the constituents of mineral deposits. Some of the volatile constituents (e. g., water, oxygen, and carbon dioxide) appear to have followed circuitous paths from the magmas, to the hydrosphere, and then to mineral deposits, but their ultimate source appears to be the magmas.

Magma crystallization may be analyzed in three stages: (1) initial, (2) intermediate, and (3) final. Mineral deposits fall, likewise, into one of these categories, depending upon what stage of crystallization is currently applicable. Thus, many metallic oxides, sulfides, and native metals are representative of the initial stage of magma crystallization. Other minerals, generally silicic, are representative of the intermediate stage of magma crystallization. Still other minerals, highly silicic, are representative of the final stage of magma crystallization.

During the initial stage of magma crystallization, the abstraction of the early crystallization products leaves a normally silicic fluid enriched in volatiles. These tend to collect in the upper part of the magma chamber until a path of preferential escape develops, permitting them, under suitable conditions, to form contact-matasomatic mineral deposits.

Toward the close of the magma solidification, some of the highly silicic accumulated fluids may pass through paths of preferential escape to form pegmatites. These fluids are accompanied or followed by aqueous solutions containing the compounds from which the ores are deposited by replacing the pegmatitic minerals. Upon final consolidation of the magma, the residual fluids are ejected along paths of preferential flow (i. e., of negative pressure gradient). From these, nearly all epigenetic mineral deposits are formed.

As the fluid cools, the gaseous constituents form magmatic liquids. These, upon nearing the surface, form hydrothermal solutions, either with or without the addition of meteoric waters. Some alteration occurs as these solutions traverse paths of preferential flow through the various wall rocks. Many become the alkaline solutions from which ores are deposited.

The paths of preferential flow may be cracks, joints, bedding planes, rock pores, and other openings that might have been produced by the geometrical, mechanical, thermal, and chemical behavior of the Earth's tectonosphere during the past 4.6 b. y. The great variety of wall rocks involved in the various conditions of temperature and pressure determines the exact locations for the hypothermal, mesothermal, and epithermal deposits.

Subsequent processes may alter the initial deposits. For example, weathering may release minerals for transport to sedimentary basins. Organic processes contribute to other alterations, including the conversion of plants and animals to fossil fuels. Some subaerial oxidation processes convert ore minerals into more usable forms, others change them into less usable forms. Metals dissolved within the oxidation zone may be transported for reprecipitation below the water table. Metals removed from above thus serve to motivate a supergene enrichment within the upper part of the sulfide zone.

Metamorphism is a cyclical or episodic process that has served to change, during the past 4.6 b. y., the form and texture of minerals, including the creation and destruction of economic deposits. Changes in temperature and pressure, or in the quantity of water present, serve to stabilize some minerals, to unstabilize others. Typical results are recombinations and recrystallizations.

Barriers within a conduit tend to influence mineral deposition. These barriers may be of many types. Gougy, impervious, pre-mineral cross-fractures may precipitate cross-breaking seams. Wherever breccia-forming walls grade into gouge-forming walls, the change in the fault-filling coinciding with the contact may form an equally effective barrier and thereby confine the ore-bearing solutions to more competent layers, e. g., when a vein passes from quartzite to shale.

The greatest concentrations of ore are usually just below gently dipping barriers, except that, wherever the solutions have had a marked horizontal component, the ore concentrates on either the hanging-wall or foot-wall side of the barrier. The actual angle deflecting the flow within the barrier determines whether the shoot "bottoms" or "tops" there.

These and other factors relating to the flow of heat and material through the tectonosphere must be included as constraints in any backward extrapolation of the effects of intrusive and extrusive activity on the origin, evolution, and present characteristics of gold deposits.

Other analyses related to the origin, evolution, and present characteristics of the paths of preferential flow within the Earth's tectonosphere include those by Bostrom et al. (1972), Dandurand et al. (1972), Fein (1973), Gill (1973), Griggs (1972), Hackett and Bischoff (1973), Heirtzler (1972), Murray (1970), Nitsan (1973), Roy et al. (1972), Steinhorn (1973), and Zonenshain (1972).

The Relationship between Intrusive and Extrusive Magma and the Low-Velocity Zone as Interpreted by the Tectonospheric Earth Model.

The low-velocity zone is related to intrusive and extrusive activity, according to the Tectonospheric Earth Model, through a common mechanism, i. e., the same driving mechanism motivates (1) magmatic activity and (2) the low-velocity zone, possibly produced by incipient melting in the upper mantle (See, e. g., Lambert and Wyllie, 1970). The postulated incipient melting is caused by the same mechanism that melts the magma that ascends to produce intrusive and extrusive activity. Thus, the low-velocity zone may be considered a complex global network, or system, of horizontal "intrusions" between asthenospheric plates and shells (Tatsch, 1972a: chapter 6). In these regions, the unbalanced force vector on some protomagma is horizontal (i. e., this may occur wherever there are insufficient vertical conduits to convey all the protomagma upward). Magma unable to proceed upward will proceed sideward, if there is a horizontal path, or it will remain in situ, if there are no paths. Intrusive and extrusive activity is closely related, therefore, to both the low-velocity zone and to the paths of preferential flow.

Other studies in related areas have been made by Dimitriyev (1972), Heirtzler (1972), Scheidegger (1972), Shtreys and Tseysler (1972), York and Helmberger (1973), and Zapletal (1968).

The Association between Intrusive and Extrusive Magmatism and Tectonic Activity as Interpreted by the Tectonospheric Earth Model.

Basaltic and rhyolitic magmatism is associated with crustal thinning which, in turn, may be caused by (1) crustal tension, (2) heating from below, or (3) a combination of these. Because the amount of crustal tension and heating in a given area at a given time is dependent upon contributions from the geometrical, mechanical, thermal, and chemical factors of the subsurface environment, it may be said that intrusive and extrusive magmatism forms a temporal spectrum, or continuum, of compositions ranging from ultra-basicity to ultra-acidity. This compositional continuum repeats in a cyclical but aperiodic manner, according to the Tectonospheric Earth Model. This accounts, according to this concept, for the great variety of basic-to-acid rocks that have been emplaced on the surface of the Earth during the past 4.6 billion years (b. y.) (Tatsch, 1972a).

Other analyses in related area include those by Azhgirei (1968), Bain (1968), Benerjee and Ghosh (1972), Bickford and Van Schmus (1973), Borrello (1972), Cepeda (1973), Corliss (1973), Dixon and Pereira (1973), Gilluly (1971, 1972), Harrison (1972), Heirtzler (1972), Hutchinson and Hodder (1972), Kaula (1972), Khain (1972), McElhinny (1972), McGetchin et al. (1973), McKenzie (1972), Miyashiro (1972, 1973), Sillitoe (1972a, 1972b, 1972c, 1973), Spall (1972), Subhotin et al. (1972), Winterer (1973), and Wright and McCurry (1973).

Intrusive and Extrusive Activity Associated with Uplifts and Thrusts as Interpreted by the Tectonospheric Earth Model.

Many mountain ranges are the product primarily of vertical uplift, rather than of horizontal crustal compression as was once thought (See, e. g., Eardley, 1963). The Laramide Rockies, for example, east of the Paleozoic miogeosyncline, were formed primarily by intrusive activity on a large scale consisting essentially of oval or irregularly broad shapes having structural relief of a few hundred meters (e. g., the Bowdoin Dome) to 12 km (e. g., the Wind River Uplift). Later Tertiary faulting modified considerably some of these Laramide "intrusives". Younger Laramides were modified by erosion, sedimentation, and subsequent intrusive and extrusive activity. When the thrust faults of the Laramide province are charted, they prove to be, for the most part, marginal to the uplifts or intrusives. There is general correlation between the amount of uplift and the type and prominence of border thrusts, suggesting a common causal mechanism. This association is very pronounced in cases of large uplifts (i. e., 6 km or more) but they may be almost undetectable in intrusives of low and intermediate uplift.

These and similar relationships between uplifts and thrusts suggest that vertical uplift (or intrusion from below) was the primary deformation and that thrusting was a secondary lateral deformation caused by unbalanced force vectors acting vertically and obliquely downward, i. e., such as would be evidenced by gravity sliding, flowing, and large-scale erosion, for example (Tatsch, 1972a).

Many mineral deposits appear to be associated with (1) the volcanism of intermediate and silicic types related to uplifts and graben faulting; and (2) the flood-basalt extrusions related to massive crustal warping along future rift margins (See, e. g., Baker et al., 1972).

Kimberlites are usually associated with intrusive carbonate (carbonatite), alkali basalts (lamprophyres), or other unusual highly alkalic rocks in regions commonly associated with epeirogenic uplift (Wyllie, 1967, 1969a, 1969b). The genesis of these rocks appears to involve the localization of volatiles or partial melting in the upper mantle. Because this genesis apparently reflects long-lived, deep-seated mantle behavior, the spatio-temporal relationships of kimberlite petrology and mineralogy is of interest to the origin, evolution, and accumulation of other mantle-derived minerals.

Very few proposals exist for the origin of the volatiles that appear to be involved not only in kimberlite emplacement but also in epeirogenic uplift (See, e. g., McGetchin et al., 1973; Tatsch, 1972a).

Related analyses include some by Damon (1972), Faure (1972), Hills (1972), and Trofimov (1972).

Subduction, Obduction, and Other Omniductive Activity as Contributions to the Evolution of Mineral Deposits during the Past 4.6 Billion Years.

Many minerals seem to arrive at the Earth's surface as a result of sea-floor spreading, crustal plate motion, continental drift, and other omniductive movements involving lithospheric plates and the magmas associated therewith. It is well to consider briefly how these geophenomena are interpreted as constituting essential phenomena in the evolution of mineral deposits.

When viewed on a global scale and over a long period of time, sea-floor spreading and other forms of crustal rifting appear to represent the surficial behavior of the Earth in those areas where the Earth's internal behavior causes crustal tension. Crustal plate motion, subsurface plate and block motion, and continental drift, in a similar manner, represent the surficial behavior of the Earth in those areas where the resultant of the Earth's internal behavior is representable as a tangential unbalanced force vector acting on a plate, on a block, or on a continent to produce motion of that block, plate, or continent relative to the bulk of the Earth or to a datum therein (Tatsch, 1972a).

In short, sea-floor spreading, crustal plate motion, subsurface plate and block motion, and continental drift, according to the Tectonospheric Earth Model, are surficial adjustments to compensate for the disequilibrated and ever-changing behavior of the Earth's interior. The Earth's surficial behavior is a temporarily delayed attempt of the Earth's surface to adjust itself to the ceaseless equilibrating behavior of its interior. In some cases a block may be moved to effect equilibration; in others, a plate may be moved. In still other cases, a rift may be formed to effect the equilibration. Sometimes, worldwide mountain ranges grow from these rifts in a single orogeny. At other times, an earthquake occurs when a block or plate adjusts itself to a more nearly equilibrated position in the never-ending attempt of the Earth to assume a condition of minimum energy. These and related matters regarding the Earth's behavior are discussed in greater detail in Tatsch (1972a).

The Evolution of the Mineral Deposits during the Past 4.6 Billion Years according to the Tectonospheric Earth Model.

When the Tectonospheric Earth Model is used to analyze the evolution of the mineral deposits during the past 4.6 b. y., it is convenient, because of the inherent geometrical, mechanical, thermal, and chemical constraints of the model, to consider three aspects of the evolution of the mineral deposits: (1) their most probable evolutionary behavior during the period 4.6 b. y. to 3.6 b. y. ago; (2) their most probable evolutionary behavior during the period 3.6 b. y. to 0.6 b. y. ago; and (3) their most probable evolutionary behavior during the period 0.6 b. y. ago to the present. This is done in the following sections.

The Evolution of the Mineral Deposits during the Period 4.6 to 3.6 Billion Years Ago.

For the purpose of this analysis, it is assumed, for reasons stated above, that the Earth's surface 4.6 b. y. ago was devoid of any surficial features other than those that would have been genetic to an accretionary process of planetary formation from the solar nebula at that time (Tatsch, 1972a: chapter 1). The evolutionary behavior of the mineral deposits during the period 4.6 b. y. to 3.6 b. y. ago would have been motivated by the Earth's internal behavior plus the influence of extraterrestrial factors operating upon the Earth during that period.

According to the Tectonospheric Earth Model, the fragments of the five "terrestrial" octants of Earth Prime became accreted onto a 5400-km primordial Earth during the period 4.6 b. y. to 3.6 b. y.

ago. Present evidence does not define clearly exactly when this accretion occurred within this one-billion-year span. Nor is it necessary to this discussion that this be known. Suffice it here to say merely that the evolution of the mineral deposits during the period 4.6 b. y. to 3.6 b. y. ago depended primarily upon two basic aspects of the Earth's behavior during that period: (1) the mineral "mix" within the accreting fragments of Earth Prime was not as homogeneous as was that of the surface of the 5400-km primordial Earth; and (2) the primordial Earth, as it accreted to its present size, behaved in accordance with the geometrical, mechanical, thermal, and chemical aspects described in Chapters 2 and 3.

It is of interest to recall, in this connection, that many of these accreting fragments bore a marked resemblance to the rocks that now comprise the Moon, the asteroids, and the meteorites (Tatsch, 1972a: chapter 19; 1974a).

The above-described behavior produced the following constraints upon the evolutionary behavior of the mineral deposits during the period 4.6 b. y. to 3.6 b. y. ago: (1) the surface of the 5400-km primordial Earth was sialic material cut by three mutually-orthogonal belts of extruding mafic and ultramafic material; and (2) the accreting material (from the fragments of the five terrestrial octants of Earth Prime) comprised a complete spectrum of material, ranging from (a) fragments of iron, nickel-iron, and other metals that might have been contained within the molten and plastic parts of Earth Prime prior to its fragmentation, to (b) sialic material (from the surface of Earth Prime) that was similar to that of the surface of the 5400-km primordial Earth.

These constraints produced the following evolutionary behavior within the Earth's surface during the period 4.6 b. y. to 3.6 b. y. ago: (1) the initial sialic minerals on the surface of the primordial 5400-km Earth and on the erstwhile surface of Earth Prime were thoroughly metamorphosed by the effects of the accretionary activity and of the geometrical, mechanical, thermal, and chemical behavior of the primordial Earth as it grew from about 5400 km to its present size; and (2) as the size of the primordial Earth grew (at an average rate of about one millimeter per year), the geometrical, mechanical, thermal, and chemical effects of the Earth's internal behavior were felt mainly through the primordial orogenic belts of the Earth during the period 4.6 b. y. to 3.6 b. y. ago, i. e., through the wedge-belts of activity of the Tectonospheric Earth Model.

This evolutionary behavior within the Earth's surface (as it grew at an average rate of about 3 micrometers per day) produced the following effects upon the distribution patterns of the mineral deposits during that period: (1) the igneous, metamorphic, and sedimentary processes associated with the mineral deposits tended to form three mutually-orthogonal belts upon the surface of the Earth; and (2), as the Earth grew, at a rate of a few micrometers per day, it tended to continually equilibrate its condition to one of minimum energy by modifying the upper few micrometers of its surface each day, through the geometrical, mechanical, thermal, and chemical aspects of its driving mechanism (Chapters 2 and 3), thereby redistributing, metamorphosing, and otherwise affecting the mineral deposits to the extent that would be expected as a result of the geometrical, mechanical, thermal, and chemical effects of the basic driving mechanism under these conditions.

The most probable distribution pattern of the mineral deposits at the end of this period, i. e., 3.6 b. y. ago, may be summarized: (1) the youngest mineral belts (i. e., those 3.6 b. y. old on the present time scale) were aligned along various segments of three mutually-orthogonal belts; and (2) the older mineral belts consisted of "fossil" belts that had been overprinted, metamorphosed, and otherwise modified to the extent that would be expected as a result of the geometrical, mechanical, thermal, and chemical effects of the Earth's internal behavior during the period 4.6 b. y. to 3.6 b. y. ago.

The Evolution of the Mineral Deposits during the Period 3.6 to 0.6 Billion Years Ago.

For the purpose of this analysis, it is assumed, for reasons previously stated, that the Earth's surface 3.6 b. y. ago was that which would be expected from the evolutionary behavior described in the previous section. It is assumed further that, because most extra-terrestrially induced effects upon the Earth ceased prior to 3.6 b. y. ago, the Earth's behavior during the past 3.6 b. y. was internally motivated (Tatsch, 1972a). The evolutionary behavior of the mineral deposits during the period 3.6 b. y. to 0.6 b. y. ago would have been motivated almost entirely by the Earth's internal behavior during that period.

This behavior produced the following constraints upon the evolutionary behavior of the mineral deposits during the period 3.6 b. y. to 0.6 b. y. ago: (1) the surface of the Earth 3.6 b. y. ago consisted of three mutually-orthogonal orogenic belts plus modified older "fossils" as described in the previous section; and (2) new activity occurred primarily within the wedge-belts of activity associated with the Earth during the period 3.6 b. y. to 0.6 b. y. ago.

These constraints produced the following evolutionary behavior within the Earth's surface during the period 3.6 b. y. to 0.6 b. y. ago: (1) the younger mineral belts (i. e., those that are now 3.6 b. y. old) were metamorphosed by the effects of the geometrical, mechanical, thermal, and chemical behavior of the Earth during this period; and (2) some of the older "fossil" belts (i. e., those older than 3.6 b. y.) were modified by uplift, erosion, metamorphism, overprinting, and by other effects expected from the geometrical, mechanical, thermal, and chemical behavior of the Earth during that period.

This evolutionary behavior within the Earth's surface produced the following effects upon the distribution patterns of the mineral deposits during the period 3.6 b. y. to 0.6 b. y. ago: (1) the igneous, metamorphic, and sedimentary processes associated with the mineral deposits tended to form three mutually-orthogonal belts upon the surface of the Earth at any time during that period; and (2) the Earth tended to continually equilibrate its condition to one of minimum energy through the geometrical, mechanical, thermal, and chemical aspects of its driving mechanism (Chapters 2 and 3), thereby redistributing, metamorphosing, and otherwise affecting the mineral deposits to the extent that would be expected from the geometrical, mechanical, thermal, and chemical effects of the Earth's basic driving mechanism under these conditions.

The most probable distribution pattern of the mineral deposits at the end of this period, i. e., 0.6 b. y. ago, may be summarized: (1) the youngest mineral belts (i. e., those of the Upper Precambrian) were aligned along various segments of three mutually-orthogonal belts; and (2) the older mineral belts consisted of "fossil" belts that had been overprinted, metamorphosed, and otherwise modified as would be expected to result from the geometrical, mechanical, thermal, and chemical effects of the Earth's internal behavior during the period 3.6 b. y. to 0.6 b. y. ago.

The Evolution of the Mineral Deposits during the Period 0.6 Billion Years Ago to the Present.

For the purpose of this analysis, it is assumed, for reasons previously stated, that the Earth's surface at the end of the Precambrian was that which would be expected from the evolutionary behavior described in the previous section. It is assumed further that the Earth's behavior during the Phanerozoic was basically internally motivated (Tatsch, 1972a). The evolutionary behavior of the mineral deposits during the Phanerozoic have been motivated, under this concept, almost entirely by the Earth's internal behavior during this period.

This behavior produced the following constraints upon the evolutionary behavior of the mineral deposits during the Phanerozoic: (1) the surface of the Earth at the end of the Precambrian consisted of three mutually-orthogonal orogenic belts plus modified older "fossils" as described in the previous section; and (2) new activity has occurred primarily within the wedge-belts of activity associated with the Earth during the Phanerozoic.

These constraints have produced the following evolutionary behavior within the Earth's surface during the Phanerozoic: (1) the younger mineral belts (i. e., those of latest Precambrian age) were metamorphosed by the effects of the geometrical, mechanical, thermal, and chemical behavior of the Earth during the Phanerozoic; and (2) some of the older "fossil" belts (Precambrian) were modified by uplift, erosion, metamorphism, overprinting, and by such other effects that might be expected from the geometrical, mechanical, thermal, and chemical behavior of the Earth during this period.

This evolutionary behavior within the Earth's surface has produced the following effects upon the distribution patterns of the mineral deposits during the Phanerozoic: (1) the igneous, metamorphic, and sedimentary processes associated with the mineral deposits have tended to form three mutually-orthogonal belts upon the surface of the Earth at any time during the Phanerozoic; and (2) the Earth has tended to continually equilibrate its condition to one of minimum energy through the geometrical, mechanical, thermal, and chemical aspects of its driving mechanism (See Chapters 2 and 3), thereby redistributing, metamorphosing, and otherwise affecting the mineral deposits to the extent that would be expected from the geometrical, mechanical, thermal, and chemical effects of the Earth's basic driving mechanism under these conditions.

The most probable distribution pattern of the mineral deposits at the present time, according to this concept, may be summarized: (1) the youngest mineral belts (for example, the Tertiary copper belts and the Tertiary petroleum belts) should be aligned along various segments of three mutually-orthogonal belts; and (2) the older mineral belts consist of "fossil" belts that have been overprinted, metamorphosed, and otherwise modified as would be expected to result from the geometrical, mechanical, thermal, and chemical aspects of the Earth's internal behavior during the Phanerozoic.

Chapter 6
CENOZOIC SEISMOTECTONOMAGMATIC BELTS AND THE ASSOCIATED GOLD DEPOSITS

The gold deposits associated with the Cenozoic seismotectonomagmatic belts comprise, according to the Tectonospheric Earth Model concept, part of the Earth's "original" early-Archean gold that has been "reworked" by subsequent seismotectonomagmatic-belt activity during the Archean, the Proterozoic, and the Phanerozoic, including the Cenozoic. It is not too surprising that the Earth's known Cenozoic gold deposits are found primarily along the Pacific "ring of fire" and along the Mediterranean-Indonesian belt (Rozhkov, 1968). The Pacific "ring of fire" essentially comprises major portions of three of the Earth's 12 Cenozoic quadrispherical arcs: Bengal-to-Aleutians, Aleutians-to-Galapagos, and Galapagos-to-Bouvet. The Mediterranean-Indonesian belt essentially comprises major portions of two others: the Gibraltar-to-Bengal and the Bengal-to-Kermadecs quadrispherical arcs.

The Origin, Evolution, and Present Characteristics of the Earth's Surficial Features: A Summary.

Basic to the seismotectonomagmatic-belt concept of the Tectonospheric Earth Model is that the Earth's present surficial features are a consequence of the Earth's internal behavior during the 4.6 b. y. that the Earth is believed to have been in existence. The salient aspects of the correlations between the Earth's surface and its interior have been discussed in Chapter 3. It is well here to look more closely at those aspects of the Earth's internal behavior that have contributed to the origin, evolution, and present characteristics of the Earth's gold deposits. To do this, it is necessary to look at both the continents and the oceans.

The continents comprise roughly 1/3 of the Earth's present

surficial area. Continental sedimentary thicknesses vary widely throughout the Earth. In a given region, these thicknesses range from an average of a few km to a maximum of about 15 km. Under these sedimentary thicknesses are roughly 20 km of metamorphic rocks (gneisses and granites). Outcrops of these occur throughout the sedimentary thicknesses in all continents. A deeper basaltic layer lies beneath the crustal metamorphic rocks. A seismic velocity anomaly (the Mohorovicic discontinuity) exists at the top of this basaltic layer. The total crustal thickness ranges from about 25 km to about 75 km.

The oceans, in contrast to the continents, are underlain by a relatively uniform basaltic crust, averaging roughly 10 to 12 km (See, e. g., Jackson et al., 1972; Reyment, 1973; Tatsch, 1972a: chapter 15; 1975d). Unlike the orogenic-cratonic structure of the continents, that of the oceans seems to be simpler and younger (See, e. g., Tatsch, 1972a: chapter 7; 1975d). It is well to summarize some of the salient aspects of the differences that exist between the continents and the oceans:

a. About 8% of the Earth's surface (i. e., roughly 25% of the present continental area) comprises very old, highly-metamorphosed Precambrian rocks. Rocks of this type are rarely found in the Oceans. However, the observed paucity of Precambrian rocks in the oceans could be the results of incomplete observational data for the oceans; and the possibility has not been eliminated that Precambrian (and Paleozoic) sediments may exist today within Layer 3 of the oceanic crust (See, e. g., Brown, 1974; Rona and Dietz, 1974; Tatsch, 1972a: chapter 15).

b. At various times, the shield areas have undergone gentle tilting and associated submersion along the affected edges.

c. At other times, the Earth's internal behavior has caused large-scale, vertical (epeirogenic) movements that have affected some shield areas, causing more-general submersion. Other shields have undergone little or no epeirogenic subsiding.

d. Occasionally, crustal rifting has separated some of the oldest shield areas. Other shields have remained more-or-less unbroken ever since the Precambrian.

An analysis of the origin, evolution, and present characteristics of the Earth's gold deposits must consider the effects of these and related horizontal and vertical movements that have affected the oceans and the continents while the gold deposits were forming.

The Earth's Gold Deposits Associated with the Cenozoic Seismotectonomagmatic Belts: An Overview.

The present distribution patterns of the Earth's gold deposits are identifiable, according to the seismotectonomagmatic-belt concept, with the surficial manifestations of the Earth's internal behavior associated with the 12 quadrispherical arcs of the Tectonospheric Earth Model. The present configuration of these 12 arcs may be listed, together with some of the applicable areas that are known to contain gold deposits:

a. Bengal-to-Kermadecs. Australia, Bangladesh, Brunei, Burma, Indonesia (including West Irian, Java, Kalimantan, and Sumatra), the Khmer Republic, Malaysia, Papua New Guinea, southern Philippines, South Viet Nam, Sri Lanka, Thailand, and the associated offshore and sea-bottom areas.

b. Kermadecs-to-Galapagos. None are known to exist,

but some may exist in association with sea-bottom nodules and on unexplored islands.

c. Galapagos-to-Gibraltar. Central America, northern South America, northwestern Africa, Spain, and the associated offshore and sea-bottom areas.

d. Gibraltar-to-Bengal. Spain, northern Africa, southern Eurasia, the Middle East, Iran, Pakistan, India, and the associated offshore and sea-bottom areas.

e. Bengal-to-Aleutians. Indonesia, Malaysia, southeastern Asia, northeastern Asia, and the associated offshore and sea-bottom areas, including the South China Sea, the East China Sea, the Sea of Japan, the Sea of Okhotsk, and the Bering Sea.

f. Aleutians-to-Galapagos. Southern Alaska, western Canada, western United States, western Mexico, Central America, and the offshore and sea-bottom areas associated with this arc.

g. Galapagos-to-Bouvet. Western South America, eastern Argentina, and the associated offshore and sea-bottom areas.

h. Bouvet-to-Bengal. Southern Africa, southeastern Africa, southern India, southeastern India, and the associated offshore and sea-bottom areas.

i. Kermadecs-to-Aleutians. Southern Alaska, various islands between Alaska and the Tonga-Kermadecs complex, and the associated offshore and sea-bottom areas.

j. Aleutians-to-Gibraltar. Western and northern Alaska,

Canadian Arctic islands, western and southern Greenland, western Europe, and the associated offshore and sea-bottom areas.

k. Gibraltar-to-Bouvet. Western Africa and the associated offshore and sea-bottom areas.

l. Bouvet-to-Kermadecs. Antarctica, New Zealand, the Tonga-Kermadecs complex, the associated islands, and the associated offshore and sea-bottom areas.

In considering these listings, it is suggested that the reader use a globe with 3 rubber bands to represent the surficial traces of the 12 quadrispherical arcs of the model. When properly placed on the globe, the 3 rubber bands will form 6 orthogonal intersections (i. e., quadruple points) associated with the 6 points of the model: Aleutians, Galapagos, Gibraltar, Bengal, Kermadecs, and Bouvet (Chapter 2). In doing this, it is well to recall (from Chapter 2) that the rubber bands represent lines that are roughly 1000 km beneath the surface of the Earth, whereas the known gold deposits are at (or near) the surface. Because the projections of heat, volatiles, and other forms of energy through 1000 km of heterogeneous material, over a period of 4.6 b. y., is not a straight-line function (See Chapter 2), the projections must be made realistically, i. e., in terms of the Earth's geometrical, mechanical, thermal, and chemical heterogeneities (See, e. g., Tatsch, 1972a).

It is necessary to recall that the above listings, insofar as they pertain to the origin, evolution, and present characteristics of gold deposits, are based on the first-order outputs of the Tectonospheric Earth Model. These outputs have a computed nominal probable error of 1571 km (Tatsch, 1972a: chapter 6). Resolutions smaller than 1571 km are not intended when making first-order analyses such as this. The resolutions obtainable with this model when using its higher-order outputs are one, two, and three orders of magnitude better (i. e., 157.1 km, 15.71 km, and 1.571 km), depending upon the amount of time and funding allocated to a given area.

No adjustments have been made in the above listings to correct the first-order predictions of the model for the differential horizontal movements that have occurred, between the Earth's present surficial features and the subtectonospheric fracture system of the model (Tatsch, 1972a: chapter 6), since the time that the gold deposits were emplaced in their present locations.

In the higher-order analyses, these adjustments, amounting to translations of as much as 140 km (and appropriate rotations) for some Cenozoic gold deposits, are made for the regions being analyzed. More-complex adjustments must be made in analyses involving "older" gold deposits. These adjustments may amount to 300 km for Mesozoic deposits (Chapter 7) and 500 km for Paleozoic deposits (Chapter 8).

The important point regarding these adjustments is that the Tectonopsheric Earth Model provides a method for determining them to any degree of accuracy, limited only by the amount of time and funding available. Also, because of the "limit stop" aspects of the orogenic-cratonic evolution of the continents (Tatsch, 1972a: chapter 7), the <u>sum</u> of these adjustments rarely exceeds 3500 km, except that adjustments for some of the oldest Archean gold deposits could amount to as much as 4900 km under the seismotectonomagmatic-belt concept of the Tectonospheric Earth Model. No <u>single</u> adustment under this concept amounts to more than a few km, because individual adjustments are computed and applied incrementally in the entire gamut of the spatio-temporal frame spanning 4.6 b. y. to a depth of 1000 km.

<u>Representative Cenozoic Gold Deposits</u>.

In order to determine how well today's actual gold deposits follow the patterns predicted by the Tectonospheric Earth Model, it is well to consider the locations and characteristics of representative examples of today's known Cenozoic gold deposits.

Adelong (Australia).

See Chapter 7.

Agatovskoye (Northeastern USSR).

The Agatovskoye gold-sphalerite-galena deposit belongs to the volcanic class of hydrothermal deposits (Nayborodin and Goncharov, 1970). The area is underlain by volcanic rocks of Cretaceous and Paleogene age. These are associated with linear Maltano-Olskaya depressions in the southern part of the Okhotsk-Chukot volcanic belt. The vein mineralization is controlled by northwest and north-south trending tear faults. The ore minerals are galena, sphalerite, chalcopyrite, pyrite, and marcasite, plus some pyrrhotite, arsenopyrite, tetrahedrite, cubanite, argentite, pyrargite, and gold. Gangue minerals are mainly quartz, calcite, and dolomite. The silver-to-gold ratio is about 5. The veins contain about 15% sulfides and are of 4 types: (a) quartz, (b) carbonate-quartz, (c) sulfide-carbonate quartz, and (d) sulfide-carbonate. Five stages of mineralization are distinguishable: (a) metacolloidal quartz, (b) gold-sphalerite-galena-quartz, (c) gold-markasite-calcite, (d) sulfur-carbonate, and (e) barite-opal-chalcedony.

Alaska

See Chapter 7.

Alaskan Lodes.

These lodes come mainly from the Juneau district, from two mines on Chichagoff Island (Chichagoff and Hearst Chichagoff), and from the Willow Creek district, north of Seward. Some minor lodes occur in the Fairbanks and Cook Inlet regions. These deposits, probably originally of Archean age, owe their present form to reworkings from seismotectonomagmatic belts that were active during the Proterozoic and the Phanerozoic. The most recent of this activity includes the alteration of Triassic volcanic rocks by intrusive sills and masses of gabbro. These, in turn, were intruded by Coast Range granodiorite and by albite-diorite during the Cretaceous. The invaded rocks were profoundly altered, after which mineralizing solutions penetrated a zone of weakness in greenstone and slate. The resulting stringers and shoots contain the auriferous quartz. Some silver, base-metal sulfides, and ankerite occur with the gold. Other details are included in the literature (See, e. g., Bateman, 1951, and subsequent references).

Alaskan Placers.

These placers include stream, beach, buried, and mixed gravels. The most productive areas are Fairbanks, Upper Yukon, and Seward Peninsula. These placers were undoubtedly an outgrowth of the seismotectonomagmatic-belt activity that developed the Alaskan lodes (q. v.). In many respects, these placers resemble those of California (q. v.). Fairbanks has produced more than 7 million oz; Nome, more than 3.5 million oz. (Brobst and Pratt, 1973).

Aldan Placers (USSR).

These placers are sililar to those at Cripple Creek. Greater details are included in the literature (See, e. g., Lamey, 1966).

American River (California).

See Mother Lode.

Amur (USSR).

The Amur deposits appear to be related to Tertiary volcanism associated wtth seismotectonomagmatic-belt activity. The variations in the shape and texture of the native gold reflects the changing conditions of mineralization related to that activity. More specifically, the variations of mineralization in surficial layers can be related to changes in the ore content, especially of gold and quartz, during the process of ore formation (Petrovskaya et al., 1971). These changes are related, supposedly, to variations in terrestrial heat flow as the paths of preferential heat flow changed (Tatsch, 1972a, 1973a).

Antiquia (Colombia).

During the 17th century, Colombia produced more than a third of the world's gold (See, e. g., Lamey, 1966). Both lode and placer deposits occur. Present production from this mine and from Caldas and Golima is less than 1% of the world total.

Argonaut (California).

See Mother Lode.

Arizona (USA).

Nearly all of Arizona's lode gold and much of its placer gold has come from deposits within a northeast trending belt known as the mountain region (Koschmann and Bergendahl, 1968). This belt is about 100 km wide and borders on the southwest margin of the Colorado Plateau. Within the belt, the deformed Precambrian rocks have been tilted, intensely faulted, and intruded by masses of ingeous rocks, such as would be expected within an active seismotectonomagmatic belt.

The deposits of Arizona range from Precambrian to Tertiary age. The porphyries at Jerome are of Precambrian age. The San Francisco and Kofa deposits occur in epithermal quartz-calcite veins of

Tertiary age. The Dos Cabezas, Gila City, Harquahala, Vulture, Wallapai, and Weaver deposits represent mesothermal sulfide-quartz veins of Mesozoic or Tertiary age. Replacement deposits occur near granite intrusives in the Agua Fria, Banner, Big Bug, Tombstone, and Turquoise areas. Gold associated with copper porphyries, of Mesozoic and Tertiary age, occurs at Ajo, Bisbee, Clifton-Morenci, Globe-Miami, Ray, San Manuel, and Superior.

Aurora (Nevada).

Nearly all the exposed rocks of the Aurora district are of volcanic origin. They comprise biotite-quartz latite, andesite, rhyolite, and basalt. Periods of erosion followed each episode of andesitic, rhyolitic, and basaltic extrusion. The oldest flow (andesite) was extruded onto a granitic basement. The ore deposits occur in veins cutting the biotite-quartz latite and andesite. Most veins strike northeast and dip steeply southeast. Numerous vein branches are based on the veins. The bulk of the veins are fine-grained quartz. The ore is quartz, adularia, argentiferous tetrahedrite, pyrite, chalcopyrite, and a mix of gold and silver with selenium. The richest ores contain free gold. The absence of tellurium is unusual. Other details are included in the literature (See, e. g., Bateman, 1951; Koschmann and Bergendahl, 1968; and subsequent references).

Australasia.

Australasia ranks among the top gold producing areas of the world. The main sources are Western Australia (q. v.), Victoria (q. v.), New South Wales (q. v.), Queensland (q. v.), New Zealand (q. v.), and New Guinea (q. v.).

Ballarat Placers.

See Victoria.

Bingham (Utah).

The Bingham (West Mountain) district, roughly 30 km southwest of Salt Lake City, is on the eastern slope of the Oquirrh Range. Most of the gold is a byproduct of copper ore. The predominant bedrock of the district is the Oquirrh group of Pennsylvanian-age limestone and sandstone more than 6 km thick. Part of ahe district is underlain by Precambrian rocks. These are intruded by several granite, granite porphyry, and monzonite stocks of Tertiary age and by monzonite sills and dikes. Some Tertiary volcanic flows, breccias, agglomerates, and ash lie along the eastern edge of the district. A series of northwest-trending folds is cut by (1) several west-to-northwest-trending thrust faults and high-angle normal and reverse faults, and (2) a system of northeast-trending high-angle reverse faults. The stocks and mineral deposits appear to be controlled by the intersections of the northeast and northwest trending fault systems. Other detiils are in the literature (See, e. g., Koshchmann and Bergerdahl, 1968, and subsequent references).

Bodie (California).

The Bodie mining district, of Mono County, California, lies in the western boundary of the Basin and Range province of eastern California. In this area, many precious metal deposits are associated with volcanic rocks of Tertiary age. Most of these are epithermal gold and silver deposits in quartz veins. The surrounding rock is hydrothermally altered. The host rocks were emplaced about 9 m. y. ago (See, e. g., Silberman et al., 1972). Volcanism ceased after ore deposition (which lasted about 1.5 m. y.) and did not resume until about 5 m. y. ago.

Boise Basin (Idaho)

The young placers of the Boise Basin are probably derived from

from the lodes of the area (See Chapter 7). The placers resemble the young placers of other parts of the world; e. g., California (q. v.).

Bolivia Placers (South America).

The young placers of Bolivia lie along the east flank of the Andes. They resemble the young placers of other parts of the world; e.,g., California (q. v.).

Bulofo River.

See New Guinea.

Bunker Hill (California).

See Mother Lode.

Caldas (Colombia).

Like Antiquia (q. v.).

California (USA).

The most productive gold-bearing area of California is the western slope of the central part of the Sierra Nevada. This region contains the Tertiary channel gravels, the Quaternary stream deposits, the Grass-Valley, Nevada-City lode district, the Aleghany and Downieville districts and the complex vein system of the Mother Lode, East Belt, and West Belt. The geology of all these areas is more-or-less the same (See, e. g., Koschmann and Bergerdahl, 1968).

The auriferous gravels of California are of two types: (1) buried placers of Tertiary age, and (2) normal stream placers of Quaternary age. The gold came from the many gold-bearing veins, including those of the Mother Lode, that had formed from seismo-tectonomagmatic-belt activity beginning during the late Jurassic. This activity comprised (1) the formation of the mountains during the late Jurassic, and (2) the intrusion of the Sierra Nevada granitic batholith during the late Cretaceous. Extended erosion during the Tertiary served to almost level the mountains, with the result that the gold from the eroded veins became concentrated in stream beds. Most of the known deposits are in the following counties: Amador, Calaveras, El Dorado, Nevada, Placer, Tuolumne, and Yuba.

At the end of the Tertiary, much of the area was covered by primarily andesitic volcanism. The old drainage system was destroyed, but a new one developed in conformity with the uplifted topography of the Sierra Nevada. During the Quaternary, the streams eroded deeply through the volcanics, exposed and eroded some of the old Tertiary channels, and reconcentrated some of the gold in gravels of the Quaternary streams. These include the American, Feather, Merced, Mokelumne, and Yuba Rivers.

The oldest rock unit, of Paleozoic age, comprises slates and greenstones that have been intruded by several episodes of seismotectonomagmatic-belt activity, primarily during the late Jurassic and late Cretaceous times. The ore bodies are of several types, the most productive being mineralized greenstones and mineralized schists. The greenstones comprise albite, ankerite, quartz, and sericite, plus some pyrite and arsenopyrite. The gold is associated with the sulfides and the quartz. The schist ore comprises ankerite plus albite, pyrite, quartz, and sericite. Free gold is associated with the pyries.

The main mining districts of California are concentrated along a 200-km, northwest-trending belt parallel to the coast inland from San Francisco. Closer scrutiny shows, however, that some gold mining is associated with extensions of this belt to the northern and southern borders of California. This belt is a surficial manitestation of a segment of the Cenozoic Aleutians-to-Galapagos seismotectonomagmatic belt. The details of some of the mines aree in other parts of this book. The details of other mines are adequately covered in the literature (See, e. g., Koschmann and Bergerdahl, 1968; Brobst and Pratt, 1973). Suffice it here to say merely that the origin, evolution, and present characteristics of the California gold deposits are consistent with the seismotectonomagmatic-belt concept. That is, the "original" Archean gold was "reworked" during the seismotectonomagmatic-belt activity that occurred in Proterozoic and Phanerozoic times. Most of the evidence of the Proterozoic and early Phanerozoic activity lies hidden or has been obliterated by subsequent Phanerozoic seismotectonomagmatic-belt activity, particularly during the late Jurassic and late Cretaceous.

Camp Floyd (Utah).

The Camp Floyd (Mercur) district, in the southern part of the Oquirrh Mountains, is unlike other major mining districts of Utah. It is primarily a gold district with silver and mercury as byproducts. The major ore deposits of the district are bedded replacement deposits of gold and gold-mercury. Silver and silver-lead replacement deposits and gold and gold-mercury fissure veins are of subordinate importance. The principal mineralization of the auriferous replacement deposits are cinnabar, orpiment, pyrite, and realgar. The gangue comprises jasperoid plus some barite and calcite. the gold is too fine-grained to be seen. Its mode of occurrence is not known, but there appears to be some relationship with carbon.

Carlin (Nevada).

Carlin represents a new type of deposit which comprises disseminated gold in bedrock. Most deposits of this type are of relatively recent discovery, dating from about 40 years ago to younger. Cortez, which is of the Carlin type, was discovered in 1966. The

Carlin deposit comprises very fine-grained gold (less than 10 microns) disseminated in silty and carbonaceous dolomitic limestone. The host rock is usually early Paleozoic (e. g., Carlin and Cortez are in a Silurian host). The disseminated gold is accompanied by silica, some pyrite and other sulfides, and barite. The ore appears to have been formed by hydrothermal replacement of the host rock. The yield is about 0.3 oz/tn plus silver running about 10% that of the gold. Characteristic trace elements include arsenic, antimony, mercury, and sometimes tungsten.

Most of the Carlin-type deposits appear to have been produced during seismotectonomagmatic-belt activity of the late Cretaceous and the Tertiary. Part of the belt extends into Idaho. Specific details of the geologic environment of these deposits is not understood. But there is a marked geochemical similarity between these deposits and some modern hot-spring deposits in certain parts of the world. The hot springs of New Zealand contain as much as 85 ppm (roughly 2.5 oz/tn) of gold and roughly 500 ppm of silver (See, e. g., Weissberg, 1969).

Deposits that resemble Carlin include Cortez, Gold Acres, and Getchell (See, e. g., Brobst and Pratt, 1973; Dickson et al., 1975; Joralemon, 1975; Roberts et al., 1971; Wells and Mueller, 1973). An initial analysis, using the seismotectonomagmatic-belt concept indicates that all continents contain some highly inviting prospects of gold disseminated in late Mesozoic and Cenozoic rocks. Most of these prospects are untested, but they include such well-known vestiges of seismotectonomagmatic-belt activity as the Antler orogenic belt of Nevada (Tatsch, 1975e).

<u>Catron County</u> (New Mexico).

The main gold district of Catron County is Mogollon. The mountain range, in which the ore deposits are found, comprises a section of 2.5-km-thick Tertiary lavas and pyroclastics. These are interbedded with sandstone and conglomerate. The rocks are cut by a network of normal faults of several ages. The ore deposits occur with argentiferous and auriferous quartz veins along the faults. The primary metallic minerals within the veins are bornite, chalcocite, chalcopyrite, galena, pyrite, sphalerite, and stromeyerite. See also Nex Mexico.

Central Eureka (California).

See Mother Lode.

Chinkuashih (Taiwan).

This copper-gold mine resembles that of Butte (Folinsbee et al., 1972). The ore is primarily within fissure veins of hydrothermally altered argillized dacite intrusive of the Pleistocene and within massive calcareous sandstones adjacent to dacite. High-grade gold ore is found within breccia pipes. Erosion of other pipes and weathered ore provided some rich placers. Chinkuashih appears considerably younger than Butte and other American Cordilleran deposits. It appears that the intrusion, cooling, fracturing, alteration, mineralization, and erosion all occurred within a period of a few m. y. Geothermal and fumarolic activity in the vicinity appear to be a later phase of the same seismotectonomagmatic-belt activity that produced the Chinkuashih deposit.

Chukotka (Eastern Siberia).

These placers, in the Margadan Province, are similar to those at Cripple-Creek (q. v.). Zagruzina (1970) has dated the associated granite at 83 to 110 m. y.

Colombia Placers (South America).

The young placers of Colombia lie along the east flank of the Andes. They resemble the young placers of other parts of the world;

e. g., California (q. v.). Production in Colombia is declining (less than 200,00 oz in 1969) but these placers once contributed a significant amount of gold.

Colorado (USA).

The gold deposits of Colorado are mostly in a northeast-trending belt, called the Colorado mineral belt, within the mountainous western part of the State. The Cripple Creek district and the following 9 minor districts lie in two southeast-trending belts that abut the northeast-trending belt in Gunnison County: Bonanza, Chalk Creek, Creede, Gold-Brick/Quartz-Creek, Lake City, Monarch, Rosita Hilla, Summitville, and Tincup.

The mineral belt of Colorado trends obliquely to the mountain ranges. This suggests that the mineral belt and the mountain ranges are vestiges of at least two different episodes of seismotectonomagmatic-belt activity. The mineral belt crosees the San Juan Mountains in the southwest part of the State. It crosses the Front, Gore-Ten Mile, Mosquito, and Sawatch ranges in the central and north-central parts of the State.

The northeast-trending mineral belt is coincident with a belt of intrusive stocks, dikes, and sills of porphyritic acidic igneous rocks that were emplaced during a late Cretaceous or early Tertiary episode of seismotectonomagmatic-belt activity. Gold, silver, and lead-zinc deposits occur, throughlut this belt, in rocks of various types, dating from the Precambrian to the Tertiary.

Closer scrutiny of the gold deposits of Colorado shows that their origin, evolution, and present characteristics are consistent with the seismotectonomagmatic-belt concept. That is, the "original" Archean gold was "reworked" during seismotectonomagmatic-belt activity that occurred in Proterozoic and Phanerozoic times. Most of the evidence of the Proterozoic and early Phanerozoic activity lies hidden or has been obliterated by subsequent Phanerozoic seismotectonomagmatic-belt activity. The following examples may be cited:

a. In the Front Range, the gold deposits are within metamorphic rocks of Precambrian and Tertiary ages.

b. In the Gore-Tenmile, Mosquito, and Sawatch ranges, the gold deposits are in sedimentary rocks of Paleozoic age.

c. In the San Juan Mountains, the gold deposits are primarily within volcanic rocks of Tertiary age.

Closer scrutiny of these and other examples from Colorado suggests the following:

a. At least 3 episodes of seismotectonomagmatic-belt activity were associated with the origin, evolution, and present characteristics of the State's gold deposits.

b. To produce the abutment pattern involving Cripple Creek and the 9 minor districts mentioned earlier, it is possible that two mutually-orthogonal seismotectonomagmatic belts were involved in the origin, evolution, and present characteristics of Colorado's gold deposits. Several alternative explanations using only one seismotectonomagmatic belt are possible. One of these possibilities involves a 90° rotation of Colarado's surface with respect to the Earth's tectonosphere. Of these possibilities, the two-orthogonal-belt concept is the more probable.

Comstock Lode (Nevada).

The oldest rocks of the Comstock Lode district are folded and faulted limestone, schist, shale, quartzite, and metavolvanic rocks of late Paleozoic or early Mesozoic age. These are intruded by late Jurassic quartz monzonite and covered by a series of Tertiary volcanics. Some of these are further intruded by Miocene diorite. The last intrusion was followed by normal faulting. The Comstock Lode and other veins were deposited along the faults. Later Pliocene or Pleistocene movement produced new faults plus some displacement along older faults.

The lode is a complex, Y-shaped shear zone roughly 4 km long and 100 m wide. The ore minerals are argentite, native gold, and stephanite plus some base-metal sulfides (See Lamey, 1966; Koschmann and Bergendahl, 1968, for other details).

Cortez (Nevada).

See Carlin, which resembles this deposit.

Cresson (Colorado).

This is part of the Cripple Creek complex (q. v.).

Cripple Creek (Colorado).

These gold deposits, associated with recent volcanogenic belts, are near-surface gold-silver formations. They are characterized by a spatial association with subvolcanic intrusions of liparites, formed at the final stages of seismotectonomagmatic-belt development and having a more-complex petrographic composition than do the effusives comprising these belts. Some deposits occur near the centers of volcanic activity. The ore bodies are formed by banded chalcedony comb quartz, adularia, and carbonates. Wall-rock alterations included propylitization, alunitization, and the development of metasomatic quartzites and quartz-adularia metasomatites.

This complex comprises almost a hundred mines, including Cresson, Golden Cycle, Portland, Stratton-Independence, and Vindicator. Other details are included in the literature (See, e. g., Brobst and Pratt, 1973; Lamey, 1966; and Stemprock, 1968). See also Guanajuato.

<u>Deadwood/Two-Bit</u>.

See Chapter 9.

<u>Delamar</u> (Nevada).

Delamar lies on the west slope of Meadow Valley Range. The district contains a thick section of tilted and faulted Cambrian sedimentary rocks. These are overlain by Tertiary andesites, latites, rhyolites, and tuff that are faulted but less tilted than are the Cambrian rocks. Small diorite bodies and basaltic-rhyolitic dikes and sills cut the sedimentary rocks. The rhyolite dikes appear to be related to the flows. The mineralization was emplaced during the volcanic activity. The major deposits occur in the oldest rocks (late Precambrian or early Cambrian). Three types of deposits are found: (1) quartzite breccia, cemented and partly replaced by vuggy fine-grained quartz containing comb quartz and sulfides within the vugs; (2) small veins of fine-grained quartz containing free gold and sulfides; and (3) bedded quartz with gold within some fractures and along bedding planes. Other deposits are in volcanic breccia, intruded by rhyolite dikes, as at Magnolia and in several limestone units scattered with silver.

A closer analysis reveals that the gold deposit at Delamar, like those of other Nevada districts, have developed in a manner consistent with the predictions of the seismotectonomagmatic-belt concept. That is, some of the "original" Archean gold of the Earth's tectonosphere was "reworked" during seismotectonomagmatic-belt activity that occurred in Proterozoic and Phanerozoic times. Most of the evidence of the Proterozoic and early Phanerozoic activity lies hidden or has been obliterated by subsequent Phanerozoic seismotectonomagmatic-belt activity. But one vestige of the Proterozoic seismotectonomagmatic-belt activity remains in the present-day auriferous Prospect Mountain quartzite.

Ecuador Placers (South America).

The young pacers of Ecuador lie along the east flank of the Andes. They resemble the young placers of other parts of the world; e. g., California (q. v.).

Elk City (Idaho).

The ore bodies of the Elk City district, a few km north of the Salmon River, occur within a system of fissure veins. This system is within gneiss and schist adjacent to the Idaho batholith. The veins extend radially from the curved contact of the intrusive granodiorite (Shenon and Reed, 1942). Banded gneiss is the most abundant rock. These and associated rocks appear to be derived from sedimentary rocks and quartzite of the Precambrian Belt series. The regional structure has a northwest trend. The veins appear to occupy crosscutting joints. The vein system cuts all rocks of pre-Cenozoic age. The richest ores tend to concentrate in zones of intense fracturing within the quartz. The ore shoots contain gold and sulfides throughlut most of the quartz.

El Oro (Mexico).

The epithermal bonanza deposit at El Oro resembles Guanajuato (q. v.).

Fairbanks (Alaska).

These gold-silver placers are of two basic types: (1) residual and hillside, and (2) stream, with the latter more common. The stream placers include bench and buried ancient stream deposits as well as recent deposits. Most of the richest placers are of the ancient stream type.

Feather River (California).

See Mother Lode.

Folsom (California).

See Mother Lode.

Fremont (California).

See Mother Lode.

Garden/Maitlands (South Dakota).

See Chapter 9.

Getchell (Nevada).

The Getchell gold mineralization is part of a north-trending belt about 50 km long. The gold mineralization commonly apexes from 10 to 30 m below the surface, regardless of whether the apex is under a hill or a swale. Not all observers agree on the exact age of the mineralization (See, e. g., Joralemon, 1975), but all indications are that it formed at a very shallow depth. One unusual feature of the Getchell ore is its extreme vugginess, with some cavities almost a meter across. The unconsolidated sediments in the hanging wall of one vein were partially silicified into a vuggy porous mass resembling a fumarolic deposit.

Other deposits like Getchell include Carlin (q. v.), Cortez, and Gold Acres (Roberts et al., 1971; Brobst and Pratt, 1973).

Gold Acres (Nevada).

See Carlin, which this deposit resembles.

Golden Cycle (Colorado).

Golden Cycle is part of the Cripple Creek complex (q. v.).

Goldfield (Nevada).

The Goldfield district is on the south rim of a desert basin that is the southern extension of a much larger basin. The oldest rocks are Cambrian flinty shale and quartzite. These have been intruded by late Mesozoic alaskite and granite. Overlying these is a thick series of Tertiary volcanics and lake deposits that has been warped into a dome and faulted.

Most of the ore deposits occur in prelacustrine rocks where they are localized in silicified zones of small complex dacite shears. Other deposits occur in andesite and rhyolite. The ores are complex sulfides, such as pyrite, bismuthinite, gold-fieldite, plus famatinite. Native gold is associated with these. Other details are in the literature (See, e. g., Koschmann and Bergendahl, 1968; Brobst and Pratt, 1973; Stemprok, 1968). See also Guanajuato.

Golima (Colombia).

Like Antiquia (q. v.).

Grant County (New Mexico).

Grant County, in southwestern New Mexico, is the richest mineral producing area of the State. The ore deposits are diverse and have yielded copper, gold, iron, lead, manganese, molybdenum, silver, and zinc. The major districts are Central, Pinos Altos, and Steeple Rock.

The Central district, which includes Fierro, Bayard, Hanover, and Santa Rita, is underlain by Paleozoic sedimentary rocks. These are overlain by Cretaceous sedimentary rocks. Folding, faulting, and igneous activity were associated with seismotectonomagmatic-belt episodes during the Cretaceous and the Tertiary. The igneous rocks include dikes, sills, and stocks of quartz diorite and granodiorite plus some younger quartz latite tuffs and andesite basalt flows. Some alteration followed each seismotectonomagmatic-belt-motivated igneous episode.

Ore deposits are of 3 principal types: (1) contact metamorphic, (2) veins, and (3) disseminated copper deposits in granodiorite porphyry. The gold content of the primary ore is very low, but the enriched, oxidized areas of the veins and disseminated-copper deposits yield a large amount of gold as a byproduct. See also New Mexico. Other details are in the literature (See, e. g., Koschmann and Bergendahl, 1968).

Great Barrier Island (New Zealand).

The Great Barrier subvolcanic silver-gold deposits comrpise a number of east-trending, quartz-filled, mineralized fissures (Ramsey and Kobe, 1974). The fissures transect andesitic volcanics and an unconformably overlying deposits of upper Tertiary to Quaternary age. Wall rock alternation is mainly a propylitic assemblage within the andesite, plus argillic assemblages and silicification near the veins. The most common mineral is pyrite. Others are sulfides, selenides, and native metals. The silver-to-gold ratio is 30. The mineralization, of Pliocene-Pleistocene age, appears to be associated with the late phase extrusions and intrusions of acid magma in the volcanic portion of the Kermadecs-to-Bouvet seismotectonomagmatic-belt activity.

Guanajuato (Mexico).

The epithermal bonanza deposits at Guanajuato bear a marked resemblance to similar Tertiary deposits in other parts of the world. These include the deposits at El Oro and Pachuca in Mexico and those in Nevada (e. g., Goldfield, Tonopah, and Virginia City), Colorado (e. g., Cripple Creek, Ouray, Silverton, and Telluride),

Romania, Nicaragua, the Philippines, New Zealand, and Japan.

These deposits are hydrothermal veins of quartz, carbonate minerals, barite, and fluorite containing gold or gold-telluride. Generally, the silver-to-gold ratio is large (Brobst and Pratt, 1973). Emplacement was by filling of open spaces within highly-altered Tertiary volcanics. Most emplacements were shallow (i. e., less than 50 m).

The richness of the ore varies, but generally it exceeds that of the older (Precambrian and Paleozoic) gold quartz lodes.

Hahns Peak (Colorado).

Metamorphic rocks of Precambrian age, sedimentary rocks of Mesozoic and Cenozoic age, and igneous rocks of Tertiary age have been proposed as the bedrock source of gold in the placer deposits in the vicinity of Hahns Peak, Routt County, Colorado (Antweiter et al., 1972). According to the seismotectonomagmatic-belt concept, these placers are vestiges of "original" Archean gold that has been "reworked" during the Phanerozoic. The north-trending Park Range to the east of Hahns Peak comprises mainly granites, gneisses, schists, and quartzites of Precambrian age. Associated sedimentary strata are mainly of Permian, Triassic, Jurassic, Cretaceous, Miocene, and Pliocene ages. Doming was caused by Pliocene intrusions comprising rhyolitic porphyry sills and a central stock having an inner breccia pipe almost a km in diameter. The King Solomon fault dominates the north side of Hahns Peak.

Hammonton (California).

See Mother Lode.

Hauraki (New Zealand).

The Hauraki district contains gold quartz veins in andesite. The main associated minerals are antimony, arsenic, copper, iron, lead, mercury, and zinc.

Homestake (South Dakota).

Like Cripple Creek.

Japan.

The dozen or so gold mines of Japan are mainly fissure fillings containing free gold or argentiferous gold associated with intrusions of andesitic and rhyolitic stocks and early Tertiary magmas. The ores are mostly crustified with base-metal sulfides, argentite, and stribnite. Other details are included in the literature (See, e. g., Bateman, 1951; Brobst and Pratt, 1973). See also Guanajuato.

Jarbridge (Nevada).

This district is named for the Jarbridge Mountains, in which it lies. The district is underlain by Tertiary volcanics that were extruded onto an eroded surface of Paleozoic rocks that had been intruded by granitic dikes and stocks. The Paleozoic sediments comprise folded limestone, quartzite, and shale. The intrusive rock is a hornblende biotite granodiorite of late Mesozoic age. The Tertiary volcanics comprise two series separated by an erosion surface.

The ore deposits are auriferous fissure veins within the older rhyolites. The vein systems strike northnorthwest and dip steeply eastward. The minerals are native gold, electrum, argentite, and some base-metal sulfides and other minor minerals.

Josephine County (Oregon).

See Chapter 7.

Juneau (Alaska).

Like Treadwell (q. v.).

Kainantu Goldfields.

See Papua New Guinea.

Kennedy (California).

See Mother Lode.

Keystone (California).

See Mother Lode.

Kinkwaseki (Taiwan).

The large gold-copper deposits of Kinkwaseki lie near the northeast corner of Taiwan. They are associated with necks and dikes of hornblende andesite, which intruded Tertiary shale and sandstone. The main deposit, the Motoyama, within andesite, comprises a large silicified zone traversed by a network of veins. The vein system contains quartz, much gold in the upper portions, and abundant enargite in the lower portions, plus some pyrite, barite, and calcite (Watanabe, 1942a). Other deposits, such as the Tyozin bodies, are sandstone replacements. One of these contains alunite plus quartz, barite, calcite, enargite, and lesser minerals, in which rich gold and silver are found. Another Tyozin deposit contains some interbedded coal seams. Where the ore has replaced the coal, the gold content is high.

Klamath Mountains (California and Oregon).

See Chapter 7.

Klondike (Yukon Territory).

The young placers at Klondite resemble those in other parts of the world; e. g., those in California (q. v.).

Kolyma (Eastern Siberia).

The Kolyma placers are similar to those at Cripple Creek (q. v.).

Koryako-Kamchatka (USSR).

Like Cripple Creek (q. v.).

Kuril-Kamchatka (USSR).

The gold distribution within the volcanic rocks of the Kuril-Kamchatka province is much like that of other Cenozoic auriferous volcanic rocks (Anoshin and Kepezhinskas, 1973).

La Grange (California).

See Mother Lode.

Lead (South Dakota).

See Chapter 9.

Manhattan (Nevada).

See Chapter 8.

Lincoln-Wildman-Mahoney (California).

See Mother Lode.

Mariinsk-Taiga (USSR).

The Quaternary placers of Mariinsk-Taiga provide an excellent example of how "old gold" has supplied the young placers (Yablokova and Ryzhov, 1973).

Mendeleyev Volcano (Kurile Islands).

Small amounts of gold are found within the orebody associated with Mendeleyev volcanic eruptions of 1880. The gold is associated with the mercury mineralization in the hydrothermally altered opaline rocks. The mercury mineralization has been studied in fumaroles caused by volcanic gases escaping along a ring fault of the caldera. The fumaroles developed after dacitic extrusions. Cinnabar is the predominant mineral (See, e. g., Ozerova et al., 1969). It is mostly in disseminated ore that contains iron sulfide, sphalerite, covellite, and rutile, plus some chalcocite, stibnite, and native sulfur and gold. Mercury sulfides are still forming today.

Merced River (California).

See Mother Lode.

Mokelumne River (California).

See Mother Lode.

Montezuma-Apex (California).

See Mother Lode.

Morobe Goldfields.

See Papua New Guinea.

Mother Lode (California).

The erosion and reconcentration of the Paleozoic gold lode deposits in the Mother Lode belt have produced important placers and related alluvial deposits (See, e. g., Lingren, 1911, 1933). These deposits fall primarily into two categoiies: (1) older deposits of Tertiary age, and (2) younger deposits of Quaternary age. The Tertiary placers were formed from down-cutting streams of late Cretaceous and early Tertiary times. Near the end of the Tertiary, rhyolite flows covered the auriferous gravels. Subsequent down-cutting streams reconcentrated the gold in new canyons. Andesitic tuff filled these later, to be eroded subsequently by new streams that cut deeper than a km, during the late Pliocene and early

Quaternary. This uncovered some of the Tertiary gravels as reconcentrated strands within terraces. Some of the old gravels that remained buried in former Tertiary channels have been drift mined for their gold. Two of the mines of this complex are the Hidden Treasure and the Mayflower. These have been studied by Lindgren (1911) and by later researchers.

The younger (Quaternary) placers fall into two general categories: (1) fairly small but rich deposits, and (2) fairly large but poor deposits. The first of these, which caused the California "gold rush" of 1849, were mined out within a short time, including 4 million ounces in 1852. The low-grade deposits are still worked, primarily by dredging, in places such as Oroville (Feather River), Hammonton (Yuba River), Folsom (American River), La Grange (Mokelumne River), and Snelling (Merced River).

National (Nevada).

The National district lies on the west slope of Santa Rosa Range. This part of the range comprises predominantly Tertiary basaltic lava flows. At National mine, basaltic tuffs, and lacustrine beds are overlain by latite, basaltic flows, and a rhyolite flow. Necks and dikes of the rhyolite cut all the older rocks. The ore deposits occur in fissure veins trending northerly with steep east or west dips. Most of the veins are rich in ruby silver and other argentiferous minerals. The National ore shoot contains coarse electrum irregularly distributed near the footwall of the quartz vein. Associated minerals are primarily sulfides, including stibnite, pyrite, chalcopyrite, arsenopyrite, sphalerite, and galena.

Nevada (USA).

Most of Nevada's gold is a byproduct of the argentiferous Comstock Lode. Other major gold districts include: Aurora, Delamar,

Goldfield, Jarbridge, Lynn, Manhattan, National, Potosi, Rawhide, Round Mountain, Silver Peak, and Wilson. The surficial geology reflects the activity of a late Mesozoic or early Cenozoic seismotectonomagmatic-belt, as revealed primarily in a southwest-trending belt of folding, large-scale thrusting, and granitic intrusions. Paleozoic sedimentary rocks occur throughout the State. Mesozoic sediments are exposed in the western and west-central areas. The massive Sierra Nevada batholith lies along the western edge of the State. The ages of granitic rocks within the central part of the State are not known. Volcanic and nonmarine Cenozoic sedimentary rocks are widely distributed. These rocks show the effects of normal faults that characterize the present basin-and-range topography. Ore deposits are associated with most of the rock types described above. Veins, contact-metamorphic deposits, and replacement deposits are associated with the granitic intrusive rocks and invaded rocks. Epithermal vein deposits occur within the Tertiary volcanic rocks.

The Comstock Lode and other major gold districts of Nevada are described individually in other sections of this book. Suffice it here merely to say that the Nevada gold deposits are consistent with the seismotectonomagmatic-belt concept. That is, the "original" Archean gold of the Earth's tectonosphere was "reworked" during seismotectonomagmatic-belt activity that occurred in Proterozoic and Phanerozoic times. Most of the evidence of the Proterozoic and early Phanerozoic activity lies hidden or has been obliterated by subsequent Phanerozoic seismotectonomagmatic-belt activity.

<u>New Guinaa</u> (Australasia).

Most of the New Guinea gold comes from river placers such as those of the Bulofo river.

<u>New Mexico</u> (USA).

See Chapter 9.

New South Wales (Australia).

There are many productive gold districts within New South Wales, including Adelong (q. v.), Cobar (q. v.), Hillend (q. v.), Mount Boppy (q. v.), and Wyalong (q. v.). There are six main types of deposits: (1) gold quartz fissure veins associated with granitic intrusives and carrying base-metal sulfides; (2) rich lenticular lenses; (3) fissure veins in sediments carrying arsenopyrite, scheelite, and stibnite; (4) saddle reefs like those associated with the Victoria lodes; (5) large replacement lodes along shear zones containing gold in massive sulfides of copper, iron, lead, and zinc; and (6) impregnations of auriferous pyrite in sediments.

New Zealand (Australasia).

Most of New Zealand's known deposits have been fairly well worked out. Three main areas of interest are Hauraki (q. v.), Otago (q. v.), and Thames (q. v.). See also Guanajuato, which resembles this deposit.

Nicaragua (Central America).

The epithermal bonanza deposits of Nicaragua resemble those at Guanajuato (q. v.).

Nome (Alaska).

These gold-silver placers, on the Seward peninsula, include beach placers and stream placers. The multi-level beach placers extend along the shore of the Bering Sea. The stream placers are richer.

North Star (Colorado).

This is part of the Silverton-Anastra complex (q. v.). See also Grass Valley.

Oatman (Arizona).

The Oatman district, which has produced gold from quartz veins in Tertiary volcanics, lies within a north-trending arcuate surficial feature. The area contains extensive exposures of Precambrian schists and gneisses that have been intruded, along a major fracture zone, by Cretaceous volcanics, by Cretaceous-Tertiary felsic dikes and plugs, by Laramide granitics, and by Quaternary basalts. Plutonism and volcanism represent a structurally controlled volcanogenic province that is a vestige of former seismotectonomagmatic-belt activity.

Ogura (Japan).

The Ogura mine is in Hukusima prefecture. The plagioclase-amphibolite with well-defined schistosity is intruded by irregular apophyses of sheared granodiorite. The former has been converted, along some schistosity planes, into diopside-plagioclase rocks (See, e. g., Watanabe, 1942b). The resulting rocks contain biotite, magnetite, and tourmaline. A high-grade gold ore occurs within magnetite and tourmaline concentrations.

Okhotsk-Chukotka (USSR).

The gold-silver mineralization of this 3500-km Mesozoic-Cenozoic volcanic belt is associated with the adularization resulting from potassium metasomatism (Ostrovskiy et al., 1970). This belt and its mineralization are associated with seismotectonomagmatic-belt activity that occurred in that area during the late Mesozoic and early Cenozoic. This mineralization resembles Cripple Creek (q. v.).

Old Eureka (California).

See Mother Lode.

Original Amador (California).

See Mother Lode.

Oroville (California).

See Mother Lode.

Otago (New Zealand).

Most of the Otago gold is from Tertiary placers. The lodes are rich, surficial stockworks that carry arsenopyrite, pyrite, and scheelite in addition to the gold.

Ouray (Colorado).

This is part of the San Juan complex (q. v.). See also Guanajuato, which this deposit resembles.

Pachuca (Mexico).

This epithermal bonanza resembles Guanajuato (q. v.).

Papua New Guinea (Australasia).

The relatively youthful tectonic setting of the mineral deposits in the New Guinea region was emphasized by Page and McDougall (1972). The mineralization, mainly copper-gold, generally is very closely associated with intrusive rocks that are an important feature of the New Guinea complex, especially in the Highlands. The igneous rocks associated with these deposits are of the calc-alkaline type, ranging from porphyritic granodiorite to andesite. The gold-bearing porphyries in the Morobe Goldfields are about 3.7 m. y. old and they appear to be associated with the porphyry intrusion of that age. Gold and silver mineralization in the Kainantu Goldfields appears to be at least as young as mid to late Miocene.

The origin of the gold in Papua New Guinea is interpretable in various ways. Two of these may be mentioned: (1) according to the simple plate-tectonics concept, the magmatic activity and mineralization was triggered by the interaction and collision between the Pacific plate and the Australian plate during the Miocene; and

(2) according to the seismotectonomagmatic-belt concept, the magmatic activity and mineralization was triggered by the seismotectonomagmatic-belt activity that occurred at that time along a segment of the Bengal-to-Kermadecs quadrispherical wedge-belt of acactivity (Chapter 4; Tatsch, 1972a, 1973a).

Park City (Utah).

Silver and lead are the chief commodities of the Park City district, but considerable gold, copper, and zinc are also mined. Predominant bedrock is a section of sedimentary formations that are folded, faulted, and intruded by igneous rocks and covered by volcanics. The oldest rocks are Precambrian. They are overlain by Cambrian and other Paleozoic rocks. Mesozoic sedimenatry rocks overlie these in many of the areas. During the Tertiary, folding and faulting preceded and accompanied igneous intrusions of granite, quartz monzonite, diorite, and diorite porphyry. Later faulting occurred during the late Tertiary or early Quaternary. The main structural unit is the north-plunging Park City anticline, onto which some of the smaller east-trending folds have been superimposed. The ore occurs in fissure veins with both sedimentary and igneous rocks. They are mostly bedrock limestone replacements near fissures.

Peru Placers (South America).

The young placers of Peru lie along the east flank of the Andes. They resemble the young placers of other parts of the world; e. g., California (q. v.).

Philippines (Western Pacific).

Forty years ago, the Philippines produced a fairly large amount of gold, exceeding that of South Dakota and Alaska, but most of the known deposits have been mined out now. See also Guanajuato, which resemble this deposit.

<u>Plymouth</u> (California).

See Mother Lode.

<u>Portland</u> (Colorado).

This is part of the Cripple Creek complex (q. v.).

<u>Republic</u> (Washington).

This resembles Cripple Creek (q. v.). See also Washington.

<u>Round Mountain</u> (Nevada).

Round Mountain resembles Aurora (q. v.) and produces from both placers and lodes. The oldest rocks of the district are lower Paleozoic limestone, jasper, and dark slaty schist. These intensely folded rocks were intruded during the Cretaceous by granitic magma bodies. Tertiary porphyritic rhyolite and associated lake beds overlie the folded Paleozoic rocks. Two periods of mineralization occurred. The first was just after the granitic intrusions and comprises huebnerite-bearing granite. The second, of Tertiary age, resulted in the formation of epithermal auriferous quartz veins within the rhyolite. The minerals of these veins comprise gold, auriferous pyrite, and some realgar, adularia, and alunite. Following the primary gold mineralization, additional fissures were formed. Into these, iron and manganese oxides and gold were deposited by supergene solutions. The gold in these fractures appears to have been derived from the primary auriferous pyrite.

Closer scrutiny reveals that the gold deposits of the Round Mountain district, like those of other areas of the world, have developed in a manner consistent with the predictions of the seismotectonomagmatic-belt concept. That is, some of the "original" Archean gold of the Earth's tectonosphere was "reworked" during the seismotectonomagmatic-belt activity that occurred in Proterozoic and Phanerozoic times. Much of the evidence of the Proterozoic and early Phanerozoic activity lies hidden or has been obliterated by subsequent Phanerozoic seismotectonomagmatic-belt activity. But vestiges of the earlier activity remain in the auriferous Mesozoic and Cenozoic mineralizations.

Shenandoah-Dives (Colorado).

This is part of the Silverton-Anastra complex (q. v.).

Shturmanskoye (Northeastern USSR).

See Chapter 7.

Silverton-Anastra (Colorado).

These copper-gold ores are within long, narrow Tertiary fissure veins. The mines include North Star, Shenandoah-Dives, and Sunnyside. See also Guanajuato, which this deposit resembles.

Snelling (California).

See Mother Lode.

South Dakota (USA).

See Chapter 9.

Stratton-Independence (Colorado).

This is part of the Cripple Creek complex (q. v.).

Sunnyside (Colorado).

This is part of the Silverton-Anastra complex (q. v.).

Takatama (Japan).

The auriferous quartz adularia veins at the Takatama mine,

Hukusima prefecture, demonstrate the effect of impervious strata. The content of the gold and silver is greatly increased as a result of overlying shale beds (See, e. g., Watanabe, 1942a).

Tayoltita (Mexico).

These ore deposits, in Durango near Tayoltita, are related to pulses of Tertiary volcanic and intrusive activity (Randall, 1972). The gold mineralization (electrum) is associated with quartz vein complexes that also contain copper-argentite. The gold occurs within a limit of 500 m thickness of the 4-km-thick Tertiary volcanics. The electrum occurs in dense andesite porphyry and appears to be related to the manganese and sodium within the gangue, the alteration envelopes, and the dikes. The richest ore shoots are in wide, irregular veins with complex intersections.

Thames (New Zealand).

The rich bonanzas at Thames once contained 1 to 6 ounces of electrum per pound of ore. The origin of this remarkable concentration is not known. The evidence is not clear-cut, but supergene enrichment is usually assumed.

Tintic (Utah).

The Tintic district, lying in Juab County along the Nevada border, is in the eastern part of the Great Basin. The East Tintic Mountains comprise a section of sedimentary rocks almost 10 km thick. These are complexly folded and faulted; intruded by quartz monzonite and monzonite stocks, plugs, and dikes; and covered by quartz latite and latite flows and pyroclastics. The oldest rocks are Proterozoic. The Cambrian is represented by quartzite, carbonate, and shale formations. The carbonates persist through the Paleozoic. The entire sedimentary section was folded into a series of north-trending anticlines and synclines such as would be expected from seismotectonomagmatic-belt activity. The rocks were also complexly faulted several times and were intruded by quartz monzonite and monzonite porphyry stocks, as well as by andesite, dibase, and latite dikes, plugs, and sills. Tertiary extrusives cover large areas in the south and eastern part of the East Tintic Mountains.

The ore deposits occur in limestone replacement veins and in fissure veins. These structures are spatially related to each other and to the intrusive rocks. The replacement ore zones are elongate bodies, continuous in strike but discontinuous vertically. They occupy a stratigraphic interval of about 2 km. The ore zones are continuous across faults. At some fault intersections, ore chimneys extend over 700 m vertically. The common ore minerals of the replacement deposits are argentite, enargite, galena, sphalerite, and tetrahedrite. Ore minerals of the fissure veins are mainly argentite, pyrite, chalcopyrite, sphalerite, and tetrahedrite. Both types of deposits are associated with bands of hydrothermal alteration. Native gold is a rare constituent, but some oxidized shoots contain flakes of native gold.

Tonopah (Nevada).

The Tonopah district, although primarily a silver producer, with a silver to gold ratio of 100, produced much gold during the late 19th century and early 20th century. The deposits consisted of replacement veins along narrow sheeted zones in Tertiary volcanics. Hypogene minerals comprise mainly argentite and some polybasite and stephanite within a quartz-adularia gangue. Some chalcopyrite, galena, sphalerite, and pyrite, plus some supergene sulfides, such as pyrargyrite and argentite, and native silver were also present. The gold was contained in both hypogene and supergene ores. See also Guanajuato, which resembles this deposit. Other details are in the literature (See, e. g., Lamey, 1966; Koschmann

Bergendahl, 1968; Brobst and Pratt, 1973).

<u>Transcarpathia</u> (USSR).

The gold-polymetallic deposits of the Transcarpathia area lie within the intermontane trough that separates the Carpathians from the Hungarian massif (Naumenko and Skarzhinskiy, 1970). The deposits are genetically associated with Neogene magmatic activity that occurred during an episode of seismotectonomagmatic-belt activity in that area about 15 m. y. ago. Volcanic rocks are represented by derivatives of acidic magma (e. g., liparite flows and ignimbrites) and by the basaltic layer of the upper mantle. The gold-polymetallic mineralization formed during the early stage of the Alpine geotectonic cycle of the seismotectonomagmatic-belt activity. The gold is associated with the liparite and ignimbrite. Other metals (e. g., Hg, As, Sb, Te, Bi) belong to the later stage of the cycle and are associated with andesite, basalt, dacite, and some of the liparite. The main structural control of the mineralization and of the magmatic rocks was provided by the Transcarpathian and Pannonian regional fault zones. The fault zones coincide with the metallogenic zones and are assumed to be products of the same seismotectonomagmatic-belt activity that produced the metallogenic zones (Tatsch, 1973a).

<u>Transylvania</u> (Romania).

The epithermal bonanza deposits of Transylvania resemble those at Guanajuato (q. v.).

<u>Treadwell</u> (Alaska).

These fairly low-grade lodes, which lie on Douglas Island, and contain some rich pockets, comprise primarily small, closely-spaced auriferous veins of quartz and calcite cutting through slates and diorite dikes.

Tuscarora (Nevada).

This deposit is similar to Aurora (q. v.).

Union (California).

See Mother Lode.

USSR Placers.

See Chapter 8.

Utah (USA).

Most of Utah's ore deposits lie within two northeast-trending belts that are offset left laterally about 150 km at the center, i. e., between Willow Springs and Gold Mountain. The most productive districts are Bingham, Camp Floyd, Park City, and Tintic, south and east of Salt Lake City. These districts are described in individual entires under the names of the districts.

As in other western states of the United States, the Utah ore deposits are closely related to the Tertiary stocks and small batholiths. Some gold occurs in all mineral deposits of the State except the iron and uranium-vanadium deposits. Most gold is a byproduct of base-metal ores. Bingham produces more byproduct gold than any other mine in the United States.

Victoria (Australia).

These Cenozoic gold deposits are probably derived from the Paleozoic gold lodes (Chapter 8).

Vindicator (Colorado).

This is part of the Cripple Creek complex (q. v.).

Virginia City (Montana).

The young placers at Virginia City, Montana, resemble those in other parts of the world; e. g., those in California (q. v.). Most of those at Virginia City are concentrated along Alder Gulch. The total production exceeds 2.5 million oz (Brobst and Pratt, 1973).

Virginia City (Nevada).

This epithermal bonanza at Virginia City resembles those at Guanajuato (q. v.).

Washington (USA).

Washington has many small mining districts, most of which produce from lodes. The most important mines are within the Republic and Wenatchee districts. The dominant geological feature of the Republic district is the 100-km-long Republic graben. Within the graben are narrow greenstone belts and other Permian and Triassic rocks. Most of the graben is filled with Tertiary lavas and bedded rocks. Cretaceous granodiorite underlies part of the area. Tertiary and Cretaceous quartz monzonite is exposed in some fault-controlled areas. The graben, formed early in the Tertiary, was faulted and somewhat disrupted during the late Tertiary. The ore deposits occur along the Republic fault, a high-angle fault trending north within the graben. Most ore occurs as fissure fillings within veins. The vein minerals comprise fine-grained to chalcedony quartz with stringers of very fine-grained chalcopyrite, electrum, native silver, naumanite, pyrite, and stephanite.

The main mine of the Wenatchee district is Gold King which produces from lode deposits. Bedrock in the vicinity is mainly steeply-dipping beds of Eocene sandstone. The ore deposits occur in a silicified dikelike body cut by many small quartz veins. The ore minerals are gold, silver, and some pyrite. No association is apparent in this mine between the pyrite and the precious metals. The gold occurs, in very finely-grained form, within the quartz.

Wenatchee (Washington).

See Washington.

Wilson (Nevada).

The Wilson district is underlain by quartz monzonite, similar to that in the Sierra Nevada batholith. The ore deposits are in a sheared zone crosscutting the quartz monzonite in a northwesterly direction. The quartz monzonite is cross-cut also by at least one granite porphyry dike. It is overlain by a series of pink and gray rhyolite flows. The ore comprises quartz and pyrite lenses in crushed monzonite within a belt just south of the major fault. The economic ores are gold and silver. The richest ore is within the near-surface, oxidized aone.

Wyoming (USA).

Wyoming produces from both lodes and placers. The placers are similar to the young placers of other parts of the world; e. g., those in California (q. v.). The lodes are older and are discussed in Chapter 9.

Yakutya (USSR).

These placers, in the Aldan region, resemble those at Cripple Creek (q. v.).

Yatani (Japan).

The Yatani mine contains both lead-zinc and gold-silver veins. These resulted from epithermal mineralizations associated with the seismotectonomagmatic-belt activity of the late Tertiary (Tatsch, 1973a). Briefly stated, the gold appears to have been transported primarily as gold-sulfur complexes rather than gold-chloride complexes (See, e. g., Hattori, 1975). Oxidation appears to have caused the gold to precipitate. This suggests neutral or slightly alkaline solutions to permit simultaneous precipitation of sphalerite and galena as the oxygen fugacity increased, through seismotectonomagmatic-belt activity, as the ore solution ascended. Other epithermal gold-silver deposits of Japan appear to have formed in the same manner. Most of these deposits are within northeastern Japan and have Neogene volcanic affiliations, mainly comprising volcanics and pyroclastic rocks.

The Yatami mine lies at the southwest margin of the Yonezawa basin. The area comprises Miocene formations unconformably overlying pre-Tertiary basement. Mineralization is probably late Miocene. The veins are fissure fillings. In most cases the lead-zinc veins occur separately from the gold-silver veins, but many are joined and otherwise closely associated. Major minerals are sphalerite, galena, chalcopyrite, pyrrhotite, marcasite, quartz, iron chloride, muscovite, adularia, manganese silicate, and manganese carbonate. Some arsenopyrite, stannite, electrum, and argentite also occur. Other details have been described (See, e. g., Hattoni, 1975).

Yuba River (California).

See Mother Lode.

Conclusions Regarding the Cenozoic Gold Deposits.

Close scrutiny of the Cenozoic gold deposits reveals that almost all of them appear to be associated with plate boundaries, rift valleys, extrusive and intrusive activity, metamorphism, hydrothermal activity, and other surficial manifestations of seismotectonomagmatic-belt activity that occurred within the upper 1000 km of the Earth during the Cenozoic. This may best be visualized by comparing the locations of the actual deposits with surficial manifestations of the seismotectonomagmatic belts that have been active during the Cenozoic (See, e. g., Figure 2-8, page 34). In doing this, it is well to recall that the dashed lines shown on the figure are roughly 1000 km <u>beneath</u> the surface, whereas the gold deposits are at the surface.

Because the projection of energy through 1000 km of heterogeneous material over a period of 4.6 b. y, is not a straight-line function (See Chapter 2), the projections must be made realistically in terms of the Earth's actual geometrical, mechanical, thermal, and chemical heterogeneities (See, e. g., Tatsch, 1972a). This has been done for each deposit, but the details of the individual projections are beyond the scope of this book. In most cases, the reader should be able to make these projections for himself, on the basis of the concepts contained in Chapters 2, 3, 4, and 5. In those cases where the reader finds that the projections are difficult or elusive, the author may be contacted with a request for his projections in those cases. There is no obligation for this. In fact, the author solicits requests from all readers who have questions in either interpreting or applying the new and unfamiliar concepts of this book to gold deposits of all types in all parts of the world.

Chapter 7

MESOZOIC SEISMOTECTONOMAGMATIC BELTS AND THE ASSOCIATED GOLD DEPOSITS

The gold deposits associated with the Mesozoic seismotectonomagmatic belts comprise, according to the Tectonospheric Earth Model concept, part of the Earth's "original" early-Archean gold that has been "reworked" by subsequent seismotectonomagmatic-belt activity during the Archean, the Proterozoic, and the Phanerozoic through the Mesozoic but not including any such activity during the Cenozoic. These latter deposits have been considered in Chapter 6.

It is not too surprising that the Earth's known Mesozoic gold deposits are found primarily in the Urals, in Kazakhstan, in central Asia, in North America, and in Australia (Rozhkov) 1968). These are the main areas where vestiges of Mesozoic seismotectonomagmatic-belt activity still remain today.

Representative Mesozoic Gold Deposits.

In order to determine how well today's actual gold deposits follow the patterns predicted by the Tectonospheric Earth Model, it is well to consider the locations and salient characteristics of representative examples of today's known Mesozoic gold deposits.

Adelong (Australia).

The gold of the Adelong field of New South Wales comes from gold quartz fissure veins associated with granitic intrusives and carrying base-metal sulfides.

Adycha-Indigirka (USSR).

The gold-antimony ore bodies of Adycha-Indigirka lie within the central part of the Verkhoyansk-Kolyma orogenic belt of northeastern Yakutia. They are controlled by tear faults that developed within the substructure from seismotectonomagmatic-belt activity. Besides the auriferous bodies, gold occurs also within quartz veins in the hydrothermally-altered brecciated zones. A typical deposit, Sarylakh, is in terrigenous Triassic sediments. Typical auriferous veins contain 50% to 100% stibnite associated with quartz. Less gold occurs within the quartz than within the stibnite, but the size of the gold specks (about 1 to 60 microns within the stibnite) increases within the quartz (Anasenko, 1970).

Agatovskoye (Northeastern USSR).

See Chapter 6.

Alaska (USA).

Most of the Alaskan gold is from placers. The lode gold comes from the Juneau district plus some from Prince William Sound. The primary lodes are associated with Mesozoic granites that have intruded rocks dating back to the Precambrian (Koschmann and Bergendahl, 1968). The intrusive belt extends from the Seward Peninsula to the Yukon Territory. The lode deposit on Unga Island in the Aleutians is in Tertiary andesite. The placers are widespread along nearly all major rivers, within beach sands in the Nome area, on Kodiak Island, Yakataga, Liturya Bay, and Cook Inlet.

Aldan Shield (USSR).

The quartz-feldspar metasomatites of the Aldan Shield were formed as a result of episodes of seismotectonomagmatic-belt activity that occurred during the late Archean, early Proterozoic, and Mesozoic. These metasomatities are characterized by (1) rare-metal and rare-earth mineralizations during the Archean and Proterozoic episodes, and (2) gold mineralization during the Mesozoic episode (Terentiev et al., 1971). All metasomatites are related and appear to be vestiges of the "original" Archean Earth. All are associated with zones of regional fractures such as would be expected to result from seiemotectonomagmatic-belt activity.

Amur (USSR).

These gold ores were formed in connection with the geosynclinal granitoids within miogeosynclinal depressions filled with monotonous sandy-schistose strata of terrigenous deposits. See USSR Lodes and USSR Placers.

Antiquia (Colombia).

See Chapter 6.

Arizona (USA).

See Chapter 6.

Baker County (Nevada).

Baker County contains about a dozen gold-bearing districts within the northeastern part of Oregon. These include the districts of Baker, Connor Creek, Cornucopia, Cracker Creek, Eagle Creek, Greenhorn, Lower Burnt River Valley, Mormon Basin, Rock Creek, Sparta, Sumpter, Upper Burnt River, and Virtue.

The oldest rocks of Baker County are Paleozoic. The productive veins occupy shear zones within metamorphic and granitic rocks. Most of the veins strike northeast. Vein minerals comprise pyrite, sphalerite, chalcopyrite, galena, tertahedrite, tellurides, and native gold. Both lode and placer deposits are mined. Some gold is associated with the sulfides.

Baley, or Balei (Eastern Transbaikal).

The Baley deposit is well-known for its gold mineralization. The host rocks comprise granitoids, volcanic-sedimentary formations of the basement, and the upper-Jurassic/lower-Cretaceous terrigenous sediments filling a graben (Yeleyeva and Rusinova, 1969; Gapontsev et al., 1969; Gapontsev, 1971). The tear faults are filled with quartz plus some clay and carbonates. The veins contain a low-grade sulfide and a high-grade gold mineralization. The host rocks were subjected to numerous episodes of hydrothermal alteration.

Ballarat (Victoria).

Ballarat is similar to Bendigo, but about half of the Ballarat gold comes from rich placers. The lodes are saddle-reef deposits and rich quartz bodies at the crest of an anticline. Where intersected by west-dipping strike faults, many rich auriferous ores are found. These lodes are similar to those at Bendigo (q. v.). The placers are described under Victoria Placers (q. v.). See also Chapter 8.

Boise Basin (Idaho).

The Boise Basin, about 40 km northeast of Boise, Idaho, is characterized by a zone of structural weakness, deformation, and intrusion known as the "porphyry belt" (Anderson, 1942). About 3 km wide, this belt trends N55°E and appears to be part of a belt of Tertiary deformation and intrusion that extends across the entire State. The entire area, including the Idaho batholith, suffered recurrent disturbances during the late Mesozoic and all of the Tertiary. These disturbances, such as would be expected from seismotectonomagmatic-belt activity (Chapter 4), included intrusions by several types of porphyry dikes and by ore deposition.

The lodes occur in two kinds of openings and appear to have been deposited after all but the latest of the dikes had been intruded. Mineralization comprised three stages: (1) base-metal sulfides, (2) gold and silver, and (3) small amounts of calcite. The ore shoots were formed, during the second stage, within fissures that had been filled during the first stage (Anderson, 1942). The lode output has been mainly gold plus some copper, lead, and silver.

Most evidence indicates that the lodes along the "porphyry belt" are not genetically related to the enclosing Idaho batholith. Rather, they appear to be related to later seismotectonomagmatic-belt activity that intruded a magma into that zone of structural weakness. The youngest of these magmas formed dikes and stocks. The lodes are concentrated mainly in and near Quartzburg and Grimes Pass, which appear to be the loci of maximum disturbance. Important lodes include Gold Hill, Pioneer, and Mountain Chief-Belshazzar.

Bolivia Placers (South America).

See Chapter 6.

Bralorne (Canada).

The Bralorne orebodies lie within the Bridge River district of British Columbia where long lenses of augite diorite cut Cretaceous sediments and greenstone. Some dikes of serpentine follow the diorite-sediment contact; others cut them. The orebodies are in quartz veins that cut diagonally across the diorite into the sediments and greenstone. Most of the veins strike east-west and dip northward 45° to 75°. There are some north-south cross veins. The east-west veins are displaced 30 to 120 m by several thrust faults. Ore comprises quartz and silicified aplite plus some arsenopyrite, sphalerite, galena, and other sulfides. The gold is mainly in free form.

Brewer (South Carolina).

The country rock in the Brewer mine is quartz sericite schist that has been derived from bedded tuffs of the slate belt. These have been folded into a northeasterly-trending syncline. A body of intruded granite to the north is believed to have been responsible for most of the hydrothermal alteration of the schist in the mine area. The ore deposit is a mass of silicified schist containing pyrite plus some bismuth, cassiterite, covellite, and enargite. Most of the gold is associated with the pyrite, but some free gold also occurs. A large mass of topaz comprises much of the northwestern part of the lode. The lode, weathered to about 20 m, forms a sandy mantle that encloses the native gold. This material also comprises the placer deposits at the Brewer mine.

Britannia (British Columbia).

The Britannia mines are located on the east side of Howe Sound, part of the Coast Range. Primarily copper producers, these mines produce also substantial amounts of gold, silver, zinc, and pyrite. The orebodies are in the roof pendant within the Coast Range batholith. This pendant, roughly 4 km by 12 km, comprises dark carbonaceous slate, grey slaty tuffs, and greenstone. The rocks appear to be mid Paleozoic; the Coast Range batholith, mid Mesozoic. The formation stikes roughly east-west, parallel to the long axis of the roof pendant. The gold, together with quartz and base-metal sulfides, formed during the fifth of 7 mineralization stages (Ebbutt, 1942).

Busugun-Khatynnakh (USSR).

The granitoid rocks associated with this deposit were emplaced in connection with episodes of seismotectonomagmatic-belt activity during the late Juraasic and early Cretaceous. Zagruzina (1971) used the Ar method to arrive at ages varying from 144 to 134 m. y.

Cable Mine (Montana).

The orebodies at Cable Mine are in a tabular mass of marmarized limestone, almost completely surrounded by granodiorite intrusion of the Laramide orogeny. The gold ore, in separate bodies, comprises coarse calcite carrying variable amounts of quartz, pyrite, chalcopyrite, pyrrhotite, arsenopyrite, magnetite, and gold. The auriferous orebodies are large, irregular pipes or cylindrical masses.

<u>California</u> (USA).

See Chapter 6.

<u>Carlin</u> (Nevada).

See Chapter 6.

<u>Castlemaine</u> (Victoria).

These saddle reefs are similar to those at Bendigo (q. v.) and Ballarat (q. v.).

<u>Central Aldan</u> (USSR).

Higher than normal contents of gold are found within the alteration aureoles controlled by faults in the Archean basement rocks of the Central Aldan region (Zakharov et al., 1969). The basement rocks comprise granitized biotite schists and gneisses. The metasomatites are represented by dense, fine-grained rocks, mainly potassium feldspar, carbonates, and pyrite. Gold is present within at least 3 different associations.

<u>Chukotka</u> (USSR).

These gold ares were formed in connection with the geosynclinal granitoids within miogeosynclinal depressions filled with monotonous sandy-schistose strata of terrigenous deposits.

<u>Clunes</u> (Victoria).

These saddle reefs are similar to those at Bendigo (q. v.) and Ballarat (q. v.).

<u>Cochrane Hill</u> (Nova Scotia).

Gold in the new Cochrane Hill deposit averages about 0.17 oz/tn. Total production could be 100,000 oz. The average width is about 15 m. There are 3 gold-bearing zones.

Colombia Placers (South America).
See Chapter 6.

Colorado (USA).
See Chapter 6.

Cortez (Nevada).
See Chapter 6.

Croydon (Australia).
See Queensland.

Delamar (Nevada).
See Chapter 6.

Dorn (South Carolina).
The main type of bedrock in the Dorn area is sericite schist. This is cut by a basic dike and a light-gray rock containing clusters of quartz and feldspar. The ore deposits are lenticular bodies and groups of stringers. The lenticular lodes are replacement deposits of quartz sericite plus some chalcopyrite, covellite, manganese, and pyrite. Gold appears to be associated with the sulfides. The stringers are chalcopyrite, galena, and sphalerite.

Dunolly (Australia).
See Victoria Placers.

Ecuador Placers (South America).
See Chapter 6.

Garden Gulley (Victoria).

This is part of the Bendigo complex (q. v.).

Getchell (Nevada).

See Chapter 6.

Gold Acres (Nevada).

See Chapter 6.

Grant County (New Mexico).

See Chapter 6.

Grant County (Oregon).

The geology of Grant County resembles that of the adjacent counties, i. e., Baker and Malheur. The rock ages range from the Paleozoic to Tertiary. The oldest rocks are argillite, chert, and greenstone. Roughly 3 km of upper Triassic and Jurassic rocks are exposed in the southwestern portion of the county. The Paleozoic rocks were intruded by diorite, gabbro, and peridotite. Tertiary rocks cover most of the county. The primary gold-producing districts of Grant County are Canyon Creek, Granite, Greenhorn, North Fork, Quartzburg, and Susanville.

Haile (South Carolina).

Haile is the most productive mine of the southeastern part of the United States. The host rock of the ore deposits is sericite schist of the slate belt. The rock was originally a tuff or volcanic ash. Several northwest-trending diabase dikes cut the schistosity trend of the rocks. The Haile ores are large masses of auriferous pyritized schist and auriferous quartz-pyrite bodies. These deposits are replacement bodies in two northeasterly-trending belts of schist. The mineralized rock is primarily fine-grained quartz containing much fine-grained pyrite and gold. Small amounts of pyrrhotite and sphalerite are also present.

Hedley (British Columbia).

The rocks at Hedley district are mainly Triassic. The productive area is primarily siliceous limestone and intercalated quartzite. During the late Mesozoic (or early Cretaceous), stocks of diorite and gabbro were intruded. These were followed by granodiorite and granite. The calcareous beds have been intensely metamorphosed. The ore deposits are irregular bodies enclosed in limestone. Arsenopyrite is the main sulfide, accompanied by some pyrrhotite, chalcopyrite, and sphalerite. The ore zone follows a minor fold. Other details are contained in the literature (see, e. g., Knopf, 1942; and later references).

Holkol (Korea).

The deposits of Holkol, in central Korea, may be related to the shearing of massive rock into tabular or chimnay-shaped deposits of contact metasomatic type (See, e. g.. Watanabe, 1942b). These occur along planes of preferential shearing (Tatsch, 1972a) and they are somewhat removed from the intrusive rocks. The deposits comprise typical contact minerals, such as diopside, phlogopite, ludwigite, and chondrodite, with some chalcopyrite, bornite, bismuthinite, gold, and bismuth. The deposits are worked for gold, copper, and bismuth.

Hustlers (Victoria).

This is part of the Bendigo complex (q. v.).

Idaho (USA).

The Idaho gold deposits are lodes and placers. The lode deposits are associated with the Idaho batholith, a massive Cretaceous granitic intrusion that occupies most of the central part of the State. Most of the gold is near contact with the batholith or within intruded rocks comprising sedimentary, metasedimentary, and volcanic rocks. Most of the deposits are auriferous quartz veins that contain various amounts of arsenopyrite, pyrite, galena, and sphalerite. Remnants still remain of a once-extensive cover of Tertiary basalt that buried an erosion surface. Many of the country rocks date back to the Paleozoic, and earlier

<u>Jackson County</u> (Oregon).

Schists of pre-Mesozoic age underlie the southwestern part of the county. Triassic greenstones and metasedimentary rocks underlie the western part, and Jurassic-age rocks occupy the northwestern part. The schists, greenstones, and sedimentary rocks were intruded at least twice: (1) by masses of peridotite, now altered to serpentinite, and (2) by later granitic rocks. Cretaceous marine remnants outcrop in the central part of the county. A narrow band of Tertiary sandstones crosses the county in a northwesterly-southeasterly direction. Miocene and Pliocene lavas, breccias, and tuffs in the center of the county have been invaded by small intrusive bodies of basalt and diorite. Much of the eastern part of the county is covered by Pliocene and Pleistocene lavas and cinder cones. The gold deposits are concentrated within granitic and dioritic intrusives in the western part of the county.

The main gold-producing districts of the county are the Ashland, Gold Hill, Jacksonville, and Upper Applegate districts. Other details are contained in the literature (See, e. g., Koschmann and Bergendahl, 1968).

<u>Josephine County</u> (Oregon).

Josephine County hasbeen one of the leading gold producers in Oregon. The oldest rock is late Paleozoic or early Mesozoic (probably Triassic) greenstone that underlies much of the southern part of the county in a northeasterly-trending belt. Quartz veins containing gold plus the sulfides of copper, iron, and zinc are common within the greenstone. The western part of the county is underlain by interbedded sedimentary and volcanic rocks of Jurassic age. These strike parallel to the Triassic greenstone. The sedimentary rocks are mainly black slates containing sandstone, conglomerate, and tuff. A greenstone unit is interbedded with the sedimentary rocks. Peridotite and serpentine occupy belts several km wide within the Jurassic rocks. These strike concordantly with the regional northnortheast trend. Sulfide and auriferous-quartz veins fill fractures within the Jurassic sediments and greenstone. Numerous channels, eroded into Cretaceous and older rocks, were filled with auriferous Tertiary conglomerates. Gold placers are found within the high terraces of Pleistocene age along the Rogue River. The main gold-producing districts of the county are Galice, Greenback, Grants Pass, Illinois River, Lower Applegate, and Waldo.

Kariyskoye (USSR).

See Chapter 8.

Keperveem (USSR).

The Keperveem auriferous area in the Chukot Peninsula is about 15 km in diameter. The country rocks are Triassic terrigenous sediments and early to middle Triassic bodies of gabbro-diabase (Davidenko, 1970). The sedimentary sequence is intruded by granitoids and dikes of diorite porphyry and lamprophyre similar to those at Nezhdaninskoye (q. v.). Zones of auriferous quartz veins are associated with gabbro-diabase and dikes. The earliest mineral association in the quartz veins is with carbonate, arsenopyrite, and quartz. At the hanging wall, in the middle of the veins, there are thin veinlets of quartz, galena, gold, sphalerite, and muscovite.

Gold tends to concentrate in the selvage and in the brecciated zones of quartz veins. The same episode of seismotectonomagmatic-belt activity that produced Nezhdaninskoye also produced Keperveem.

Klamath Moutains (California and Oregon).

This placer belt, which is over 300 km long, comprises residual mantle deposits, Cretaceous conglomerates, plus Tertiary and Quaternary gravels. The Cretaceous auriferous conglomerates appear to be marine shore deposits that contain fine gold and platinum in "black sand".

Klondike (Yukon Territory).

See Chapter 6.

Korea (Asia).

The Korean gold is associated with auriferous quartz fissure veins in sediments or igneous rocks dating from the Precambrian to the Mesozoic. Besides several typical contact-metamorphic deposits, most of the gold is in sulfides, plus some cinnabar, molybdenite, and wolframite.

Lake City (Colorado).

This part of the San Juan complex (q. v.).

Lebedinoye (USSR).

The Lebedinoye ore field, in central Aldan, comprises three tectonic blocks characterized by different (1) depths of basement rocks, (2) structures of the upper sections, and (3) specific magmatic and mineralization features (Safonov, 1971). Tectonic structures, developed in connection with seismotectonomagmatic-belt activity dating back to the Proterozoic, have been most instrumental in localizing the gold deposits within this field. Maximum gold concentration occurs within a narrow, elevated block characterized by laccoliths, dikes, sills, and other tectono-intrusive structures such as would be expected to result foom seismotectonomagmatic-belt activity.

Libby Quadrangle (Montana).

The ore bodies in the Libby Quadrangle are controlled mainly by the major persistent faults and by minor faults and fractures near the larger faults (Gibson, 1942). The richest ore is in the smaller faults and fractures. The gold-quartz veins occur primarily within the bedding-plane faults.

The area is underlain by sedimentary rocks of the Belt series, i. e., sandstone, argillite, and shale with a total thickness of almost 15 km. Intrusive igneous rocks, which underlie a small part of the area, are mainly Precambrian metadiorite sills plus some Mesozoic quartz monzonite stocks. During the Mesozoic, the rocks and enclosed sills were folded into broad, open anticlines and synclines trending northwesterly. The folded rocks are cut by north to northwesterly trending, steep faults.

The gold-quartz veins are generally conformable with the Belt strata. They comprise mainly gangue that is earlier than most of the sulfides and precious metals. The ore shoots are confined to areas where renewed faulting permitted mineralizing solutions to enter the veins. The ores in this area resemble those of the Clark-Fork and Murray districts of Idaho.

Loddon Valley (Australia).

See Victoria Placers.

Malheur County (Oregon).

Considerable quantities of gold have been produced from two districts of Malheur County that lie along the Jackson County border.

These are the Malheur and Mormon Basin districts. The geology of these destricts is the same as that of Jackson County (q. v.).

<u>Manhattan</u> (Nevada).

See Chapter 8.

<u>Mineral Ridge</u> (Nevada).

See Silver Peak.

<u>Montana</u> (USA).

Most of the Montana mining districts are associated with the Boulder batholith and nearby stocks and dikes of late Cretaceous or early Tertiary age. The ore deposits occur in the margins of the granitic rocks or in the wallrocks. The 4 most important districts are Butte, Helena, Marysville, and Virginia City.

Butte, one of the great copper areas of the world, produces most of the Montana gold as a byproduct of the copper ores. The Helena district includes the famous Last Chance Gulch placer deposits. The Marysville district, which includes the Silver Creek and Bald Butte areas, produces primarily from veins plus some from placers. The Virginia City district, which includes the Alder Gulch and Summit areas, produces primarily from placers plus some lodes.

<u>Morning Star Mine</u> (Victoria).

See Thompson River.

<u>Mother Lode</u> (California).

See Chapter 8.

<u>Mount Boppy</u> (Australia).

The saddle reefs at Mount Boppy, in New South Wales, are reminiscent of those at Ballarat (q. v.) and Bendigo (q. v.) and other parts of Victoria. They undoubtedly owe their existence to the same active seismotectonomagmatic belt thet was associated with the saddle reefs in Victoria.

Mount Isa (Australia).

See Queensland.

Mount Morgan (Queensland).

Mount Morgan is primarily a copper mine, but it produces about 5 dwt of gold per ton of ore. The orebody is a large beet-shaped mass about 350 m long, 200 m wide, and 300 m deep. The pyritic orebody is a stockwork in auriferous and cupriferous tuffs. See also Queensland.

Needle Mountain (Colorado).

This part of the San Juan complex (q. v.).

Nevada (USA).

See Chapter 6.

New Chum (Victoria).

This is part of the Bendigo complex (q. v.).

New Mexico (USA).

See Chapter 9.

Nezhdaninskoye (USSR).

The Nezhdaninskoye gold deposit is at the boundary of two elongate structural zones within a major synclinorium comprising Permo-Triassic and Jurassic terrigenous rocks. The hydrothermal quartz veins within diorite porphyry and lamprophyre dikes are controlled by the intersection of the regional faults (Silichev,

1970). Arsenopyrite comprises about half of the sulfides that constitute about 5% of the quartz-gold ore. Other ore minerals are pyrite, sphalerite, galena, tetrahedrite, boulangerite, jamesonite, falkmanite, and gold. According to the seismotectonomagmatic-belt concept, the deposit was formed during a late Paleozoic episode of seismotectonomagmatic activity that continued into the early Mesozoic.

<u>Northair</u> (British Columbia).

This precious-base metal deposit lies roughly 110 km north of Vancouver. The economic veins yield from 0.1 to 0.8 oz/tn of gold, from 0.8 to 14.3 oz/tn of silver, and some copper, lead, and zinc. The total gold in this deposit probably exceed 200,000 oz.

<u>Okhotsk-Chukot</u> (USSR).

See Chapter 6.

<u>O</u>. <u>K</u>. <u>Mountain</u> (British Columbia).

The O. K. Mountain deposit, about 2 km south of Rossland, is in high-grade shoots within quartz-filled fissure veins. Besides the O. K. Mountain mine, there are 3 associated mines: I. X. L., Midnight, and Golden Drip. The country rocks comprise altered andesite and basic volcanics overlying a stock of serpentine (Stevenson, 1942). All are intruded by monzonite tongues and lamprophyre dikes. The veins contain free-milling gold localized in shoots, plus some pyrite, chalcopyrite, and galena. The veins occur only in the volcanics and not in the serpentine. Many ore shoots within the veins have been localized by cross fissures. Most veins strike S80^{o}W and dip south, from 35° to 75°, toward the serpentine mass but without entering it.

<u>Oregon</u> (USA).

Most of Oregon's gold comes from two areas: (1) Josephine and Jackson counties in the southwestern part of the State, and (2) Baker, Grant, and Malheur counties about 500 km to the northeast. These deposits are discussed under individual entries devoted to these counties.

<u>Pennsylvania</u> (USA).

See Chapter 8.

<u>Peru</u> <u>Placers</u> (South America).

See Chapter 6.

<u>Potosi</u> (Nevada).

The geology of the Potosi district is fairly complex. The eastern part of thisarea is underlain by Cambrian shale, slate, and limestone and by Pennsylvanian conglomerates. The higher central part of the district comprises a Cretaceous granodiorite stock. Overlying this is a thrust plate of Ordovician cal-silicate, marble, phyllite, and recrystallized chert. North-trending normal faults dislocated the rocks after the episode of thrust faulting. The major ore deposits occur along the north-trending Getchell fault. They are sheetlike masses distributed irregularly in veins associated with the fault branches and localized at depth along the main fault. The veins comprise sheared, mineralized sedimentary rocks cut by quartz-calcite veinlets containing a soft carbonaceous mix. Most of the gold is in this mix. Barite, chabozite, gypsum, and fluorite are also present as gangue. The major ore minerals are cinabar, arsenopyrite, marcasite, pyrite, pyrrhotite, realgar, stibnite, magnetite, sheelite, and gold. The latter accurs as minute particles associated with the carbonaceous mix, the magnetie netite, the pyrite, and the marcasite, The richer gold ores are associated with abundant realgar.

Queensland (Australia).

The Queensland deposits are primarily lodes of auriferous quartz and sulfides. They include Charters Towers, Croydon, Mount Isa, Mount Morgan (q. v.), and Tympie.

Rawhide (Nevada).

Most gold of the Rawhide district comes from lodes but a significant amount comes also from placers. The richest placers are on the southeastern slope of a hill. The lode deposits occur within a network of veinlets crosscutting the country rocks of rhyolite, dacite, and andesite. The most strongly mineralized ores are kaolinized rhyolite. Ore minerals are argentite, cerargite, and electrum.

Round Mountain (Nevada).

See Chapter 6.

San Juan (Colorado)

This region includes many important mineral districts, mostly base-metal with silver and minor gold, including Bonanza, Creede, Lake City, Needle Mountain, Ouray, and Telluride. During the Paleozoic, Mesozoic, and Cenozoic, a seismotectonomagmatic belt was active in this area. Consequently, the region has a complex history of sedimentation, folding, erosion, large volcanic eruptions, intrusions, and faulting. Other details are contained in the literature (See, e. g., Bateman, 1951; and subsequent references).

Sansei (Korea).

The Sansei mine is in the remote mountains near the Yalu River in northern Korea. The hypothermal gold quartz veins are in the Precambrian injection-gneiss and granite (Kato, 1942). The first mineralization at Sansei followed the thrust planes formed in late

Mesozoic times. The source of the mineralization appears to have been the granitic magma that was widely intruded throughout the Korean peninsula during the late Mesozoic. The second mineralization deposited high-grade gold ores in the variously inclined cracks that were developed during seismotectonomagmatic-belt activity.

Scarsdale (Victoria).

These saddle reefs are similar to those at Ballarat (q. v.) and Bendigo (q. v.).

Seneca (British Columbia).

This deposit, near Harrison Mills, yields about 0.1 oz/tn of gold and 2.5 oz/tn of silver plus 14% zinc, 2.5% copper, and some lead.

Sheep Creek Camp (British Columbia).

The Sheep Creek camp comprises a 10-km-long belt south of Nelson in the Kootenay district. The rocks are Precambrian and are intruded by granitic rocks of the Nelson batholith of Mesozoic age. The sediments are folded into a tight anticline, trending north-south, overturned to the west. A large drag fold is on the western side of the anticline. The auriferous fault fissures strike northeasterly to easterly across the anticline. Fault fissures are filled with quartz, pyrite, pyrrhotite, galena, and sphalerite. Some 60 m and less displacements occur. The ore shoots lie within open fissures (Walker, 1942).

Shturmanskoye (USSR).

The Shturmanskoye deposit lies within the Che-Cheka river basin of the Yana-Kolyma orogenic system. The gold veins are located around the Khattyrnakh granite massif (Zagruzina, 1970). The veins appear to have been associated with the prebatholithic dikes that formed in connection with an episode of seismotectonomagmatic-belt activity in that area about 155 m. y. ago. The gold content of the veins is about 0.2 g/tn. The quartz fold mineralization was followed by gold-rare-earth mineralization, and then by a gold-silver mineralization connected with subvolcanic potassic igneous rocks produced by seismotectonomagmatic-belt activity during the period 52-to-58 m. y. ago.

Silver Peak (Nevada).

The rocks of the Silver Peak (Mineral Ridge) district comprise Cambrian and Ordovician limestones, slates, aand quartzites. These were intruded during a late Jurassic or early Cretaceous episode of seismotectonomagmatic-belt activity involving mainly alaskite. The gold deposits are in quartz lenses within the limestone and appear to be genetically related to the late siliceous phase of the alaskite (Koschmann and Bergendahl, 1968). Native gold is finely disseminated within the quartz. Other gold is associated with scattered sulfides in the lenses.

South Carolina (USA).

The geology of the South Carolina gold belt is similar to that of North Carolina (q. v.). The Haile (q. v.) and Brewer (q. v.) mines are in the schistose rocks of the slate belt. The Dorn (q. v.) mine is in similar rocks near the Georgia border in the southwestern part of the State.

South Verkhoyan (USSR).

These gold ores were formed in connection with the geosynclinal granitoids within miogeosynclinal depressions filled with monotonous sandy-schistose strata of terrigenous deposits.

Tarnagalla (Victoria).

See Victoria Placers.

Telluride (Colorado).

This is part of the San Juan complex (q. v.). See also Guanajuato, which resembles this deposit.

Thompson River (Victoria).

The Thompson River mine, about 130 km east of Melbourne, provides an example of the mobilization of gold and its concentration into dilation zones (See, e. g., Keays and Kirkland, 1972). The orebody occurs in a small dike that appears to have remained a closed system with respect to the gold. The dike is a classic example of ladder-reefs of auriferous quartz, such as are found at Morning Star Mine. The ores consist of chalcopyrite and pyrite, the latter having replaced pyrrhotite, plus some pentlandite, cubanite, and chalcopyrite. Gold and other precious metals appear to occur in discrete phases of the ores. The associated hornblendite dike has been extensively sheared, faulted, and hydrothermally altered through seismotectonomagmatic-belt activity (Tatsch, 1973a). The alteration and carbonization are most intense in the sheared portions of the dike. The copper, palladium, iridium, gold, and some cobalt and nickel in the main part of the dike have migrated from sheared and altered rocks into the segregated sulfides trapped along the footwall of the dike.

Within the Thompson River complex, gold shows the greatest redistribution. For example, relatively fresh hornblende, only 100 m from the orebody, contains 3% chalcopyrite but only 0.1 ppm of gold. This contrasts with the auriferous content of the Skaergaard (Greenland) samples where 3% copper sulfide ore contains twice as high a concentration of gold (Vincent and Crockett, 1960). These examples, and others, show that gold can be transported by hydrothermal solutions. Two other such examples may be cited: (1) the hot-spring precipitates at Steamboat Springs, Nevada (White, 1967), which contains 10 ppm gold; and (2) similar precipitates at Broadlands, New Zealand (Weissberg, 1969), which contain 85 ppm gold. According to some observers (See, e. g., Browne, 1969), the gold may be either magmatic in origin or leached from the country rocks. According to the seismotectonomagmatic-belt concept (Chapter 4; Tatsch, 1973a, 1975a), the gold may be derived from a linear combination of both these processes.

Transbaikalia (USSR).

These deposits are within areas of tectonic activity with completed folded. Mesozoic magmatism, with which the commercial gold mineralization is associated, is related to tectonic activity of the Baikalides and Hercynides. The granitoid intrusions of the tectonic belts occur as chains within the axial parts of trched uplifts. The Jurassic gold deposits are closely related, spatially and trmor rally, to the tectonic activity, according to the seismotectonomagmatic-belt concept.

Virginia City (Montana).

See Chapter 6.

Warren (Idaho).

The gold deposits of the Warren district, in north-central Idaho, occur in fracture zones within lenticular quartz bodies (Reed, 1942). The fractures trend about N80°E and dip sharply to the south. Post-mineral faulting has offset some of the veins. Lamprophyre dikes, intruded after ore deposition, follow some of the crosscutting, northwesterly-trending fractures. The veins contain quartz monzonite and related rocks of the Idaho batholith. Some quartzite, schist, and gneiss, probably of the Belt series of the district, crop out in the northwestern part of the district. The ore minerals are gold, galena, sphalerite, tetrahedrite, stibnite, and pyrite. These were emplaced during the latter part of the mineralization and have been localized in shoots.

Woods Point (Victoria).

The auriferous quartz veins at Woods Point occupy fractures in a dike of diorite. The lenticular dike strikes north and dips steeply westward. In many respects, the geology resembles that of Bendigo (q. v.), suggesting that the same type of seismotectonomagmatic-belt activity created both deposits.

Wyalong (New South Wales).

Like Adelong (q. v.).

Wyoming (USA).

See Chapter 6.

Yanokolyma (USSR).

These gold ores, like many others of this age in other parts of the world, were formed in connection with the geosynclinal granitoids within miogeosynclinal depressions filled with monotonous sandy-schistose strata of terrigenous deposits.

Conclusions Regarding the Mesozoic Gold Deposits.

Close scrutiny of the Mesozoic gold deposits reveals that almost all of them appear to be associated with those areas of the Earth's surface where vestiges of Mesozoic seismotectonomagmatic-belt activity remain today. These areas include some in all continents but particularly those in the Urals, in Kazakhstan, in central Asia, in North America, and in Australia.

In most cases, the reader should be able to associate each Mesozoic gold deposit with the corresponding vestiges of Mesozoic seismotectonomagmatic-belt activity on the basis of the concepts contained in Chapters 2, 3, 4, and 5. In this connection, the general comments made at the end of Chapter 6 apply also here, including the invitation for the reader to correspond with the author in those cases where either the concepts or their application to specific gold deposits is not clear to the reader.

Chapter 8

PALEOZOIC SEISMOTECTONOMAGMATIC BELTS AND THE ASSOCIATED GOLD DEPOSITS

The gold deposits associated with the Paleozoic seismotectonomagmatic belts comprise, according to the Tectonospheric Earth Model concept, part of the Earth's "original", early-Archean gold that has been reworked by subsequent seismotectonomagmatic-belt activity during the Archean, the Proterozoic, and the Paleozoic, but not during the Mesozoic nor Cenozoic. The Mesozoic and Cenozoic deposits have been considered in earlier chapters.

It is not too surprising that the Earth's Paleozoic gold deposits are found primarily in western Siberia, in eastern Sayan, in China, in eastern United States, and in Canada (Rozhkov, 1968). These are the main areas where vestiges of Paleozoic seismotectonomagmatic-belt activity still remain today.

Representative Paleozoic Gold Deposits.

In order to determine how well today's actual gold deposits follow the patterns predicted by the Tectonospheric Earth Model, it is well to consider the locations and characteristics of representative examples of today's known Paleozoic gold deposits.

Altai (USSR).

See USSR Lodes.

Appalachian (Eastern USA).

See Chapter 9.

Bald Mountain (South Dakota).

Rocks of the Bald Mountain district are mainly southwesterly-tilted strata of the Cambrian Deadwood formation, underlain by

Precambrian schist. Numerous dikes, sheets, and irregular bodies of Tertiary phonolite and rhyolite porphyry cut the older rocks. The gold deposits are in replacement bodies in the Deadwood formation. Ore deposits occur in vertical shoots, probably mineralized fractures through which the mineralizing replacemtnt solutions traveled. Two types of ore occur: (1) the primary blue ore, and (2) the oxidized red ore. Most of the ore mined is red ore, consisting mainly of pyrite, arsenopyrite, and auriferous limonite associated with the fine-grained pyrite. The blue ore, highly-refractory, is not extensively mined. Other details are in Koschmann and Bergendahl (1968) and subsequent references.

Ballarat (Victoria).

These gold ores were formed in connection with the geosynclinal granitoids within miogeosynclinal depressions filled with monotonous sandy-schistose strata of terrigenous deposits. See Victoria. See also Chapters 6 and 7.

Barkerville (British Columbia).

The Barkerville gold belt is a narrow belt of rocks in the Cariboo district. The rocks contain a multidude of auriferous quartz veins (Hanson, 1942). The Cariboo series, apparently of Precambrian age, comprises quartzite, argillaceous quartzite, argillite, and limestone, partly converted to schist and marble. The series has been folded into a broad northwesterly-trending anticline. Younger sedimentary and volcanic rocks lie on the flanks, The series is intruded by quartz porphyry dikes and sills. The gold belt lies on the northeastern limit of the anticline and has a northwesterly trend. The ore-bearing rocks are interbedded quartzite and argillite. The orebodies are steeply dipping auriferous pyritic quartz veins most of which strike northeasterly, perpendicular to the strike of the rocks.

Belei (USSR).

See USSR Lodes.

Bendigo (Australia).

These unique saddle-reef deposits of Victoria lie within Ordovician slates and sandstones that have been (1) compressed into sharp anticlines and synclines, (2) intruded by granite batholiths, and (3) deeply eroded. Fifteen parallel anticlinal lines within this complex are spaced 200 to 350 m apart. The ore deposits are saddle reefs localized at the crests of anticlines within alternating beds of slate and sandstone. Numerous spurs within the complex appear to have been the channelways for the mineralizing solutions. One saddle extends 3 km. The principal mines are New Chum, Garden Gulley, and Hustlers.

The ore comprises auriferous quartz with some pyrite, arsenopyrite, pyrrhotite, galena, sphalerite, stibnite, molybdenite, and bournonite. This complex is unusual in that the tellurides are absent. However, dolomite and ankerite are present. The primary deposition appears to have been by replacement, but some open-spaced deposition was involved.

Similar saddle reefs are found at Ballarat (q. v.), Beringa, Blackwood, Castlemaine, lunes, Scarsdale, and oth r areas of Victoria. See also Yenesei and Victoria. Other details are contained in the literature (See, e. g., Bateman, 1951; McKinstry, 1942b).

Beringa (Victoria).

These saddle reefs are similar to those at Bendigo (q. v.) and Ballarat (q. v.).

Big Bell.

See Western Australia.

Blackwood (Victoria).

These saddle reefs are similar to those at Bendigo (q. v.) and Ballarat (q. v.).

Bonanza (Colorado).

This is part of the San Juan complex (q. v.).

Burke County (North Carolina)

Most of the Burke County gold occurs within quartz veins of the South Mountain region. These mountains comprise mica-hornblende schists and gneisses. These are garnetiferous formations cut by diorite, gabbro, pegmatite, and granite. Erosive weathering has uncovered the substrata to great depths. Most of the outcrops are saprolite. The schists and gneisses are cut by small quartz veins comprising milky quartz and spent sulfides containing disseminated gold. See also North Carolina. Other details are in Koschmann and Bergendahl (1968) and subsequent references.

Cabarrus County (North Carolina).

The main deposits of Cabarrus County are in the Phoenix and Reed mines. Country rocks near the mines are mainly fine-grained volcanic tuffs that have been intruded by a greenstone till. The volcanic rocks are part of the Paleozoic slate-belt sequence. The bedrock is covered by a thick mantle of auriferous saprolite. The greenstone lodes are interlaced with auriferous quartz fissure veins. See also North Carolina.

Central Aldan (USSR).

These deposits are associated with the formation of alkaline-earth and alkaline minor intrusives localized mainly in the Cambrian carbonaceous series overlying an Archean crystalline basement. The location of these intrusions in controlled by deep faults connected with flat platform structures, such as would be expected to result from seismotectonomagmatic-belt activity in that area at various times since the Archean. See also Chapter 7.

Central Bohemia (Europe).

This auriferous area of Czechoslovakia occurs in the oldest formations of the Central Bohemian pluton comprising abyssal rocks of late Paleozoic age. Proterozoic and early Paleozoic contact-metamorphic sediments comprise biotite horfelses and spotted schists interbedded with conglomerates, augite schists, and shales. These are overlain by Silurian limestones. Both the sediments and the orthogenesis have been affected by northeasterly-striking reverse faults. The later tectonic zones, which trend from north to northwesterly, are mineralized (Moravek and Rus, 1970). Enrichment occurs where the veins cut the earlier auriferous mineralization of the plutonic formation. This suggests that the gold has been mobilized from earlier aurifierous veins, such as would have resulted from seismotectonomagmatic-belt activity in that area.

Charters Towers (Queensland).

Fissure veins in Permian granodiorite produce the gold at Charters Towers to depths of 1 km. See also Queensland.

Chatkalo-Kuraminskaya (USSR).

These gold-telluride deposits are confined to Paleozoic andesite and dacite porphyries and their tuffs. The distribution of the gold-quartz veins containing sulfides and tellurides is controlled by fault zones (Rozhkov, 1970). Gold mineralization is related to small Permian granodiorite-porphyry intrusions. The orebodies indicate intensive replacement of country rocks during mineralization. The gold tellurides characterize the final stage of ore formation. Other ore minerals include pyrite, chalcopyrite, sphalerite, galena, tetrahedrite, bismuthinite, gold, altaite, and hessite.

Cobar (New South Wales).

The deposits at Cobar are large replacement lodes lying along shear zones. They carry gold in massive sulfides of copper, iron, lead, and zinc.

Cochrane Hill (Nova Scotia).

See Chapter 7.

Colorado (USA).

See Chapter 6.

Creede (Colorado).

This is part of the San Juan complex (q. v.).

Darasun (USSR).

See USSR lodes.

Davidson County (North Carolina).

Most of the gold within Davidson County is from the Cid District. Bedrock in this district comprises bands of northeasterly-trending rocks that have been metamorphosed by seismotectonomagmatic-belt activity into slates and schists, compressed into large folds, and intruded by gabbro-diabase dikes. The ore deposits are quartz-chalcopyrite-ankerite veins oxidized to depths of 15 m. The veins contain argentiferous cerussite to depths of 20 m. Below that, they comprise galena, manganese oxides, and sphalerite. The principal mines of Davidson County are Conrad Hill and Silver Hill. For other details, see Koschmann and Bergendahl (1968) and subsequent references.

Deadwood / Two-Bit (South Dakota).

See Chapter 9.

Delamar (Nevada).

See Chapter 6.

Empire (California).

See Grass Valeey.

Fairfield Valley (North Carolina).

See Transylvania County.

Garden (Maitland).

See Chapter 9.

Gebeit (Africa).

Located in the northern Red Sea Hills, of Sudan, the Gebeit gold mine has been worked for about 3500 years. The rocks are mainly silicic metavolcanics of the Nafirdeib series, part of the basement complex of the Red Sea Hills. The main structure comprises north and northeasterly trending folds and faults. These are cut by two later sets of fractures trending west and northwesterly. The northwesterly-trending fracture system appeats to be associated with the Red Sea rift system (See, e. g., El-Boushi, 1972). Free gold occurs in quartz-pyrite-arsenopyrite veins and in the adjacent wall rock. The wall rocks are altered to chlorite, epidote, pyrite, and calcite. The gold mineralization appears to be hydrothermal activity associated with the late Proterozoic or early Paleozoic emplacement of granite during seismotectonomagmatic-belt activity in that area at that time.

Grant County (New Mexico).

See Chapter 6.

Grass Valley / Nevada City (California).

The two largest lode gold mines of California are in this district. About 80 km northwest of the Mother Lode (q. v.), these gold-quartz veins occur primarily within granodiorite and metamorphic rocks. Most veins are only a few meters thick. The ores contain native gold and pyrite sulfides. The Empire vein has been worked to a depth of over 3 km. Other details of the area are contained in Lamey (1966) and subsequent references.

Gunnar (Manitaoba).

Located about 175 km northeast of Winnipeg, the Gunnar gold mine comprises lenticular orebodies of quartz within shear zones, some sulfides, and free gold. Rock outcrops are Precambrian sedimentary and volcanic. These have been intruded by sills of gabbro and diorite. An anticlinal fold strikes just west of north and plunges steeply to the south. Unsheared ellipsoidal andesite (greenstone) comrpises the most favorable host rock. (See, e. g., Lord, 1942).

Hillend (New South Wales).

These rich lenticular lenses once produced over $3 million worth of gold from 10 tons of quartz. Other details are contained in the literature (See, e. g., Bateman, 1951; and subsequent references).

Karamazar (USSR).

Native gold occurs in some of the quartz-carbonate veins at Karamazar. The mineralization is of middle Carboniferous age. The gold appears to be associated with granite-porphyry intrusions (Koschmann and Bergendahl, 1969). But gold is absent from the younger veinlets containing galena and chalcopyrite.

Kariskoye (USSR).

The gold at Kariskoye, in eastern Transbaikal, occurs (1) in quartz-actinolite-magnetite, and (2) in quartz sulfide veins. The country rocks are gabbro-diorite, diorite, and granodiorite of Paleozoic or early Mesozoic age (Litvinov et al., 1970). Upper Jurassic rocks outcrop about 1 km from the deposit. The ore veins are closely associated with dikes of porphyry lamprophyre and granodiorite. The gold-polymetallic mineralization was the third of four stages that included also (1) quartz-tourmaline, (2) quartz-magnetite, and (3) quartz-carbonate, ranging in temperature from 535° to 50° C, with the gold stage at about 335° - 230° C. The mineralization solution appears to have flowed along the fractured zone in the southeasterly direction. All these characteristics are consistent with origin and evolution of the gold deposits during various episodes of seismotectonomagmatic-belt activity in that area at various times since the earliest Archean.

Korea (Asia).

See Chapter 7.

Lena (USSR).

See USSR Placers and USSR Lodes. Other details are contained in the references (See, e. g., Konovalov, 1973).

Lynn (Nevada).

Placer gold occurs within a wide area of the Lynn district, including Lynn, Rodeo, Sheep, and Simon creeks. A large lode was discovered at Carlin in 1962. This mine is in a window of a mountain thrust fault. The ore comprises very fine-grained gold within fractured and altered Paleozoic siltstone and limestone beneath the thrust fault. Other details are in Koschmann and Bergendahl (1968) and subsequent references.

Manhattan (Nevada).

The bedrock of the Manhattan district comprises Cambrian quartzite, limestone, and schist and Ordovician chert, slate, and quartzite. Small patches of Jurassic granite are exposed locally. Tertiary lavas, tuffs, and intrusive bodies comprise the bedrock in the northern part of the district. The Cambrian formation overthrusts the Ordovician formation. The productive deposits occur within the hanging wall of the thrust fault, primarily in Cambrian limestone and quartzose schist. The limestone orebodies display a complex assemblage of metals, including pyrite, stibnite, realgar, orpiment, cinnabar, and free gold within a gangue of calcite, quartz, fluorite, sericite, levernierite, and some adularia. The richest ores are those in the quartzose schist. They comprise networks of small quartz-adularia veins carrying pyrite and free gold. The placer gold has come primarily from deep gravels within the Manhattan Gulch.

The deposits at Manhattan, like those at other Nevada districts, has developed in a manner consistent with the predictions of the seismotectonomagmatic-belt concept. That is, some of the "original" Archean gold of the Earth's tectonosphere was reworked during the seismotectonomagmatic-belt activity that has occurred in that area during the Proterozoic and Phanerozoic. Most of the evidence of the Proterozoic and early Phanerozoic activity lies hidden or has been obliterated by subsequent Phanerozoic seismotectonomagmatic-belt activity. But one vestige of the early Phanerozoic seismotectonomagmatic-belt activity remains in the present-day auriferous Gold Hill formation of Cambrian age.

Other details are contained in the literature (See, e. g., Koschmann and Bergendahl, 1968; and subsequent references).

Maslovskiy (USSR).

The area occupied by the Maslovskiy deposit lies within the

tectonic border zone between two formations near Ivdel in the northern Urals: (1) the Tagil basin comprising middle Paleozoic lavas and sediments, and (2) the horst-anticline of Saldinsk-Verkhisetsk comprising metamorphic rocks of upper Proterozoic-Cambrian age. The faults were reactivated during further seismotectonomagmatic-belt activity during the middle Paleozoic, the upper Paleozoic, and the Neogene. The hydrothermal gold and the associated metasomatic iron deposits appear to be controlled by Devonian fault zones, up to 1/2 km wide and dipping at angles up to 90° (Deryabin, 1970). The faults are filled with hydrothermally altered breccias. The beresitized hanging walls are fairly rich in copper, zinc, cobalt, lead, arsenic, boron, gold, and silver.

According to the seismotectonomagmatic-belt concept, Maslovskiy and related northern Ural deposits contain vestiges of "original" Archean gold that was "reworked" during the Proterozoic and early Phanerozoic. Most of the effects of this seismotectonomagmatic-belt activity has been obliterated by subsequent activity. However, the Maslovskiy deposits is one example of gold that was reworked during the Devonian.

Mecklenburg County (North Carolina).

The St. Catherine and Rudisil mines, the principal gold producers in Mecklenburg County, are about one km apart, on opposite ends of the same vein. The country rocks at the Rudisil mine is granite and schist. The ore occurs within pockets of 2 quartz veins. These veins follow roughly along the granite-schist contact. They are cut by several diabase dikes. The orebodies comprise intensely-fractured pyrite and quartz located wherever the veins flatten. The St. Catherine mine is similar to the Rudisil. Other details are in Koschmann and Bergendahl (1968) and subsequent references.

Minusinsk (USSR).

See USSR Lodes.

Mother Lode (California).

Comprising both lode and placer deposits, this system of veins is a narrow belt, about 1 km wide and roughly 200 km long, on the western foothills of the Sierra Nevada. This vein system is associated with the steeply dipping greenstones, schists, slates, and some intrusive serpentine. The slates and schists were deposited as sediments during the middle Paleozoic and part of the Mesozoic. Some volcanic green schists and greenstones are associated with these. Associated intrusives range from the late Paleozoic to mid Mesozoic. During the late Mesozoic, the rocks were isoclinally folded and intruded by various types of magma, mainly granitic. This suggests that a seismotectonomagmatic belt was active in this area during the Mississippian, the Jurassic, and the Cretaceous.

Although the granites are widespread in the higher parts of the Sierra Mevada, they are scarce within the MotherLode belt, where the principal igneous rocks are serpentines derived from peridotites. The gold deposits within the Mother Lode belt appear to have formed at the termination of the late igneous activity. Most of the gold depoists are two types of mesothermal emplacements: (1) quartz veins, and (2) mineralized country rock. The quartz veins consist primarily of coarse milk-white quartz in lenticular masses. At the edges and ends, the lenticular veins grade into quartz stringers mixed with country rock. Minor amounts of ankerite and albite plus some sulfides and native gold are present. The mineralized country rock comprises auriferous greenstones and schist.

Lode gold deposits occur also in 2 subparallel belts east and west of the Mother Lode belt. The veins of the east belt are generally narrower and richer than are those of the Mother Lode belt. Other details are in Lamey (1966) and subsequent references.

Muruntau (USSR).

The Muruntau auriferous deposit of the Uzbek region occurs in the northern part of a large anticline comprising Silurian and Devonian sedimentary-metamorphic series (Yudin, 1971). The main mineralization comprises pyrite, arsenopyrite, and some pyrrhotite. The host rocks appear to have undergone extensive contact metamorphism and hydrothermal-metasomatic alteration during an episode of seismotectonomagmatic-belt activity. The host rocks contain extensive disseminated sulfide mineralization of sedimentary-metasomatic origin. Veins and stockwork gold deposits are hydrothermal in origin, apparently produced in several stages.

Nevada City (California).

See Grass Valley.

New Mexico (USA).

See Chapter 9.

Nezhdaninskoye (USSR).

See Chapter 7.

North Carolina (USA).

North Carolina has produced more gold than any other state of the northeastern part of the United States. The mines occur within 2 physiographic provinces: (1) the Piedmont, and (2) the Blue Ridge. Most of the rich deposits are in the Piedmont province of Cabarrus, Davidson, Mecklenburg, and Rowan Counties. The main deposits of the Blue Ridge province are in Burke and Transylvania counties. These deposits are described in individual descriptions devoted to these counties.

Much of the bedrock of the auriferous areas is granite, gneiss, hornblende gneiss, and mica schist of Precambrian and Paleozoic age.

These rocks were intruded by Paleozoic diorite, granite, and monzonite. Some Paleozoic intrusives of dunite, pyroxenite, and syenite are also present in both these physiographic provinces.

When the gold deposits of North Carolina are analyzed in detail, it is found that they, like those of many other areas of the Earth, have developed in a manner consistent with the predictions of the seismotectonomagmatic-belt concept. That is, in this area, some of the "original" Archean gold of the Earth's tectonosphere was "reworked" during seismotectonomagmatic-belt activity that occurred in Proterozoic and Phanerozoic times. Most of the results of the Proterozoic and early Phanerozoic activity now is hidden beneath the surface or has been obliterated by subsequent Phanerozoic seismotectonomagmatic-belt activity in North Carolina. But some vestiges of still remain within the present North Carolina gold belt and geology associated with it.

Other details are in Koschmann and Bergendahl (1968) and subsequent references.

Okhotsk (USSR).

See USSR Lodes.

Pennsylvania (USA).

Almost all of Pennsylvania's gold has been produced from the Cornwall iron mine, about 8 km south of Lebanon. The gold occurs as a byproduct of iron, together with some copper and silver. The oldest rocks of the area are Cambrian limestone and dolomite. These are overlain by Ordovician rocks. The entire Paleozoic section was folded into a recumbent northeasterly-trending synclinorium. This folding and considerable thrust faulting of the same age are such as would be expected from seismotectonomagmatic-belt activity at that time.

The Paleozoic rocks are unconformably overlain by Triassic

shales, sandstones, and conglomerates. These are warped into a northerly-dipping homocline. The easterly-trending contact between these two series is a fault. Large Triassic diabase dikes, plugs, and sills intrude the sedimentary rocks. In some places, a sill separates the Triassic and Paleozoic rocks. Orebodies of replaced limestone occur within a trough at the Triassic-Cambrian contact. The ore comprises actinolite and magnetite plus some chalcopyrite, chlorite, diopside, phlogopite, pyrite, and serpentine. Small amounts of gold occur within the copper concentrates, suggesting that the gold is associated with the chalcopyrite.

Other details are contained in the literature (See, e. g., Koschmann and Bergendahl, 1968; and subsequent references).

Potosi (Nevada).

See Chapter 6.

Putintsevskoye (USSR).

Some gold and silver occur within the upper parts of the massive and banded ore of the lead-zinc-quartz-barite type in the Putintsevskoye deposit of southeastern Rudnyy Altai (See, e. g., Mikhalevskiy, 1969).

Round Mountain (Nevada).

See Chapter 6.

Rowan County (North Carolina).

The richest gold area of Rowan County is the Gold Hill District which is about 30 km long. The district is underlain by two main types of bedrock: (1) volcanics comprising breccias, flows, slates, and tuffs, and (2) intrusive diorite, granite, and greenstone. The Gold Hill fault separates these two types of bedrock, with the volcanics within the southeastern belt and the intrusives within the northwestern half of the area. The volcanic rocks have been tightly folded near their contact with the intrusives. The intrusives range in age from the greenstone (oldest) to the granite. Most of the veins are in a zone of minor faulting within the volcanics. Quartz veins and adjacent silicified wall rock contain the richest gold deposits. The ore minerals include auriferous pyrite, chalcopyrite, galena, and sphalerite. Other details are contained in the literature (See, e. g., Koschmann and Bergendahl, 1968; and subsequent references).

Rudisil (North Carolina))

See Mecklenburg County.

St. Catherine (North Carolina).

See Mecklenburg County.

Sierra County (New Mexico).

For many years one of New Mexico's foremost mining centers, Sierra County has produced from both placers and lodes. The placers, in Sierra County's richest gold area (Hillsboro), lie in a dissected alluvial fan complex east of Animas Hills. The most productive placers are those of Luxemburg, now drained by the Greenhorn, Grayback, and Hundikori gulches. The most productive lodes are in the Animas Hills. The bedrock, mainly extrusive andesite and latite, is intruded by 2 small monzonite masses associated with dikes. Paleozoic limestone lies beneath the volcanic rocks. Most of the lode gold occurs in quartz veins lying along the andesitic-latitic dikes. The ore minerals are bornite, chalcopyrite, and pyrite, accompanied by gold and silver. Two principal accessory ores are molybdenite and tetradymite. See also New Mexico. Other details are contained in the literature (See, e. g., Koschmann and Bergendahl, 1968; and subsequent references).

Silver Peak (Nevada).

See Chapter 7.

South Dakota (USA).

See Chapter 9.

Squaw Creek (South Dakota).

The Squaw Creek district includes the Carbonate, Elk Mountain, and Ragged Top areas. The district is a plateau, roughly 8 x 4 km, on the northwestern side of the Black Hills dome. The caprock of the plateau is Mississippian limestone underlain by Cambrian sedimentary rocks that are exposed along streams cutting the limestone. The sedimentary rocks are intruded by Tertiary porphyries. In the Ragged Top area, a phonolite laccolith is intruded at the base of the Carboniferous beds. The orebodies occur within the flat-lying Mississippian limestone.

The deposits occur in silicified vertical fractures and in irregular masses of silicified Mississippian limestone. The most productive deposits, west of the Ragged Top area, are flat shoots the surface. These shoots appear to be lateral extensions from the tops of the fissure veins. The orebodies were forced to form flat by the impervious cap on the plateau. The orebodies comprise silicified limestone and fluorite containing gold, silver, and tellurium. The presence of the latter suggests that the precious metals probably occur as tellurides in unaltered ore. The veins appear to narrow at depth and to pinch out at about 20 m down.

Other details are contained in the literature (See, e. g., Koschmann and Bergendahl, 1968; and subsequent references).

Sredniy-Darsun (USSR).

Sredniy, together with Darsun (q. v.), typify the eastern Transbaikal region. This region has complex bismuth minerals, such as $Pb_2(Bi,Sb)_2S_5$ and $Pb_3(Bi,Sb)_2S_6$, associated with the gold-sulfide deposits. The formation of the accompanying lead-antimony bismuth sulfosalts occurred during the final period of the principal stage of mineralization following an episode of seismotectonomagmatic-belt activity in this area at that time.

Titagara (USSR).

See USSR Lodes.

Trans-Baikal (USSR).

See USSR Lodes.

Transylvania County (North Carolina).

Considerable quantities of auriferous gravels still remain in the Fairfield Valley placers that once produced significant quantities of gold. The gravels lie mainly within the northern drainage area of the Toxaway River. Other details are in Koschmann and Bergendahl (1968) and subsequent references.

Urals (USSR).

See USSR Placers and USSR Lodes.

USSR Lodes (USSR).

Most of the USSR lodes are in the Trans-Baikal region, including the Altai, Belei, Darasun, Minusinsk, Titagara, and Ural districts. Other smaller lodes are widely scattered throughout Siberia, in the drainage basins of the Amur, Lena, and Yenisei rivers, and in the Okhotsk regions. Most deposits are narrow quartz veins containing a fraction of an ounce of gold per ton, associated with base-metal sulfides, arsenopyrite, and some tourmaline. The associated granitic rocks were intruded at the end of the Paleozoic, during an episode of seismotectonomagmatic-belt activity in that area at that time. Other details are in Bateman (1950) and subsequent references.

<u>USSR Placers</u> (USSR).

Most of the USSR placers are in the Amur, Lena, Ural, Yenisei, and Yakut regions. The Lena placers are derived primarily from small gold stringers in folded, peneplaned Precambrian rocks. The Ural gravels are predominantly low-grade, small, and occur along the entire eastern slope of the range. The Yenisei deposits are derived from fairly rich Precambrian gold quartz veins. The USSR placers resemble the placers in other parts of the world; e. g., California (q. v.). Closer scrutiny reveals that they may be divided into fossil placers and young placers (See, e. g., Brobst and Pratt, 1973; Bateman, 1951).

<u>Victoria</u> (Australia).

Victoria is an intensely mineralized gold province, the two principal mines being Ballarat (q. v.) and Bendigo (q. v.). This lode gold is primarily native gold in quartz reefs in lower Ordovician rocks. The reefs were probably formed during the early Devonian when a seismotectonomagmatic belt was active in that area (See, e. g., Thomas, 1953). The gold-bearing sulfides appear to have been formed later, during the middle Devonian. Rich alluvial deposits formed during the Cenozoic, and some were buried by widespread Tertiary basalt flows. Other details are contained in the literature (See, e. g., Bateman, 1951; Lamey, 1966; and subsequent references). See also Victoria Placers.

<u>Victoria Placers</u> (Australia).

The Victoria placers are similar to those in California. They consist essentially of four types: (1) pre-early basalt gravels; (2) high-level gravels, (3) pre-later basalt gravels, and (4) post-later basalt gravels. The <u>pre-early basalts</u> are found at elevations of 600 m and more above present stream levels. They are overlain by sedimentation and early basalts. The high-level gravels, similar to the pre-early basalts, are probably lacustrine and floor-plain deposits. They cap and flank the hills above present stream levels. The pre-later basalt gravels comprise 3 groups: (1) below-stream level gravels underlying the younger basalt (e. g., Ballarat (q. v.) and Loddon Valley) that contain coarse gold and numerous nuggets; (2) stream gravels underlying sand and clay beds to depths of over 100 m below the present surface (e. g., Ballarat (q. v.), Dunolly, and Tarnagalla); and (3) buried coastal gravels underlying basalt or sediments. The post-later basalt gravels were formed by a late Tertiary uplift that motivated the rapid erosion of rich reefs and former placers into present-day stream bed concentrations.

The Victoria placers, according to the seismotectonomagmatic-belt concept, are a good example of how "original" Archean gold has been "reworked" by the seismotectonomagmatic belts that have been active in Victoria during the Proterozoic and Phanerozoic. Other details are in the literature (See, e. g., Bateman, 1951; and subsequent references).

Virginia (USA).

The gold-bearing areas of Virginia are east of the Blue Ridge mountains in a belt over 300 km long. The ore deposits are in northeasterly-trending schists and gneisses of sedimentary and igneous origin. These are cut by auriferous quartz-pyrite veins that are generally concordant with the dip and strike of the host rocks.

When considered in conjunction with the auriferous deposits of Pennsylvania, North Carolina, Tennessee, Georgia, and Alabama, the gold deposits of Virginia are seen to have developed in a manner consistent with the predictions of the seismotectonomagmatic-belt concept. That is, some of the "original" Archean gold of the Earth's tectonosphere was "reworked" during seismotectonomagmatic-belt activity that occurred in Proterozoic and Phanerozoic times.

The main deposits are in Farquier, Fluvanna, Goochland, Orange, and Spotsylvania counties. Two veins in Farquier County produce gold from pyrite in veins and wallrocks of the Franklin and House mines. The Tellurium, Moss, and Busby mines produce from pyrite veins and lenses along the Fluvanna-Goochland border. The main mine in Orange County is Vaucluse, which produces from quartz lenses that carry auriferous pyrites. The Whitehall mine, in Spotsylvania, produced from placers.

Other details are contained in the literature (See, e. g., Koschmann and Bergendahl, 1968; and subsequent references).

Yakut (SSSR).

See USSR Placers.

Yenisei (USSR).

See USSR Placers and USSR Lodes.

Zolotaya Gora (USSR).

This copper-gold formation in the Urals is basically a vein of diopside-garnet-chloritic composition containing some hypogene chalcocite, native copper, and bright-red auricuprite. It appears that the high copper content is associated with reworking of older copper-pyrite ores by cupriferous hydrothermal solutions.

Conclusions Regarding the Paleozoic Gold Deposits.

Not all gold deposits that were emplaced during the Paleozoic are known. Some may never be discovered; others will be found through use of advanced geological concepts and better exploration techniques. One area that probably contains unknown but discoverable gold deposits lies along the Gibraltar-to-Bengal quadrispherical arc, particularly within the Alpine geosyncline (See, e. g., Evans, 1975). An extension of the same principle suggests that the same situation pertains to the segments of the other 11 quadrispherical arcs that were active during Paleozoic times.

Close scrutiny of the known Paleozoic gold deposits reveals that almost all of them appear to be associated with those areas of the Earth's surface where vestiges of Paleozoic seismotectonomagmatic-belt activity remain today. These areas include western Siberia, eastern Sayan, China, eastern United States, and Canada.

In most cases, the reader should be able to associate each Paleozoic gold deposit with the corresponding vestiges of Paleozoic seismotectonomagmatic-belt activity on the basis of the concepts contained in Chapters 2, 3, 4, and 5. In this connection, the general comments made at the end of Chapter 6 apply also here. This includes the invitation for the reader to correspond directly with the author in those cases where either the concepts or their application to specific gold deposits is not clear to the reader.

Chapter 9

PROTEROZOIC SEISMOTECTONOMAGMATIC BELTS AND THE ASSOCIATED GOLD DEPOSITS

The gold deposits associated with the Proterozoic seismotectonomagmatic belts comprise, according to the Tectonospheric Earth Model concept, part of the Earth's "original" early-Archean gold that was "reworked" by subsequent seismotectonomagmatic-belt activity during the Archean and the Proterozoic but not during the Phanerozoic. Deposits of the Phanerozoic have been considered in earlier chapters. It is not too surprising that the Earth's Proterozoic gold deposits are found primarily within the platforms and their folded peripheries, including those of western Siberia, southern and western Africa, the United States, Canada, Brazil, India, Finland, and Australia (See, e. g., Rozhkov, 1968; Sestini, 1973). These are the main areas where vestiges of Proterozoic seismotectonomagmatic-belt activity still remain today.

In analyzing the distribution patterns of the Proterozoic gold deposits, it is well to keep in mind two things: (1) the continental areas were not necessarily the same as they are today (See, e. g., Symons, 1975); and (2) the Proterozoic lasted for 2 b. y., or over three times as long as the entire Phaneroziic and about 30 times as long as the entire Cenozoic. This, considered against a background of continuous seismotectonomagmatic-belt activity in various parts of the Earth, makes it meaningless to identify a <u>single</u> Proterozoic pattern of gold deposits. Rather, one must search for patterns within deposits that are of equal age (say within about 50 m. y.). When this is done, the Proterozoic is seen to comprise not one pattern but many, perhaps 40 or more.

<u>Representative Proterozoic Gold Deposits</u>.

In order to determine how well today's actual gold deposits

follow the patterns predicted by the Tectonospheric Earth Model, it it well to consider the locations and characteristics of representative examples of today's known Proterozoic gold deposits.

Alabama (USA).

A belt of auriferous gneisses and schists extends from Georgia into east-central Alabama. It is overlain by gently dipping unmetamorphosed sedimentary rocks of the Gulf Coastal Plain. The gneisses and schists are bordered on the northwest by the folded Paleozoic sedimentary rocks that form the southern end of the Appalachians. The gold deposits occur in the Talladega slate, the Hillabee chlorite schist, the Wedowell formation, and the Ashland mica schist. The ages of these rocks are not clearly defined (See, e. g., Koschmann and Bergendahl, 1968), but according to the seismotectonomagmatic-belt concept, the "original" Archean gold of this aeea has been "reworked" by active seismotectonomagmatic belts during the Proterozoic and the Phanerozoic. Most of the gold deposits are in Clay, Cleburne, Randolph, and Tallapoosa counties. The only significant producers are the Arbacoochee of Cleburne County and Hog Mountain of Tallapoosa County.

Aldan Shield (USSR).

See Chapter 7.

Appalachian (USA).

Most of the gold from the Appalachian region has come from placers. Lenticular gold quartz veins occur in Alabama, Georgia, and the Carolinas, within old crystalline rocks. The quartz is associated with carbonates, apatite, chlorite, albite, tourmaline, and garnet. The metallic minerals are gold, common sulfides, and molybdenite, plus some tellurides, enargite, and tetradymite. Although over a million oz of gold were produced from this region, mainly in the 19th century, the mines, such as Dahlonega, Franklin, Haile, Hogback, and Taber, are now only of historical interest.

The present placers, probably from original Archean gold, have been reworked by seismotectonomagmatic-belts that were active in that region during the Proterozoic and Phanerozoic.

Arizona Lodes (USA).

Over half of Arizona's gold comes from lodes, about 1/3 from base-metal byproducts, and the remainder from placers. Roughly 80% of the lode gold comes from deposits within 100 km of the southwestern margin of the Colorado Plateau. This belt has been subjected to intense faulting, ingeous intrusion, and such other activity as might be expected from the seismotectonomagmatic-belt episodes that have occurred there, particularly during the Proterozoic, Laramide, and Tertiary times.

The Proterozoic veins are of Lindgren's hypothermal type. They occur mainly in the Bradshaw Mountains and in the Cherry Creek district. They strike predominantly northerly and dip westerly. They appear to be genetically related to the intrusive activity associated with seismotectonomagmatic-belt activity of the Proterzoic.

Nearly half of Arizona's lode gold has come from veins in areas intruded by igneous activity during the Laramide. These mesothermal veins are localized by structural features that do not involve the Tertiary lava flows. Included are the districts of Tombstone, Congress, Octave, Harquahala, Fortuna, and Bradshaw mountains. The most productive veins strike northeasterly-southwesterly.

Veins of the Tertiary (epithermal) type are found in the Oatman, Union Pass, Kofa, and Mammouth districts. The vein fissures strike predominantly northwesterly-southeasterly. They appear to have resulted from seismotectonomagmatic-belt activity during the middle-to-late Tertiary. Roughly 100 lode gold districts occur in Arizona. See also Chapter 6.

Ashanti (Ghana)

See Chapter 10.

Bankfield (Ontario).

Bankfield is 10 km west of Little Long Lac (q. v.). The main rocks are conglomarates, graywacke, slates, and iron formation. These are intruded by dikes of albite porphyry and by irregular dioritic masses. Except for its stratigraphically more-elevated position, Bankfield resembles Little Long Lac; and these and related deposits were undoubtedly produced by the same episode of seismotectonomagmatic-belt activity.

Boliden (Sweden).

The gold-copper-arsenic deposits of Boliden are in the Skellefte mining district of Västerbotten, northern Sweden. The Precambrian rocks of the area comprise remarkably fresh, acid-to-intermediate lavas and pyroclastics. Some beds of basic lavas are conformably overlain by phyllitic sediments. About 3 km south is a granite intrusive. The rocks are steeply folded with the schistosity striking generally east-westerly, reminiscent of similar deposits in Quebec (e. g., Cadillac) and other Precambrian areas of the world.

The mineralization occurred in 3 stages: (1) arsenopyrite and some sulfides of Bi, Co, Cu, Fe, Pb, Sb, and Zn; native gold; and large amounts of apatite, quartz, and rutile; (2) intrusion of lamprophyric dikes followed by deposition of tourmaline and quartz plus sulfides and sulfosalts, tellurides, and gold in veins and lens-shaped bodies; and (3) deposition of pyrite in 2 large bodies. The mineralization solution appears to have been of the high TP type and fairly concentrated when introduced into the ore zone. Other details of the area have been discussed (See, e. g., Odman, 1942; and later references).

Brownhill-Oroya (Australia).

See Chapter 10.

Bulawayo (Rhodesia).

See Chapter 10.

Bushveld (Africa).

These complex intrusions of basic magmas with accompanying magmatic and post-magmatic deposits contain auriferous Cu-Ni-Pt ores. This deposit, like one in Sumatra, is unusual in the association of platinum with the gold.

Cadillac (Quebec).

The Cadillac complex includes the gold mines at Central Cadillac, Lapa Cadillac, O'Brien, Pandora Cadillac, and Wood Cadillac. The auriferous deposits are hypothermal quartz veins and related replacements in Archean volcanic and sedimentary rocks. Besides free gold, the metallic minerals include arsenopyrite, pyrite, pyrrhotite, and some sphalerite and galena. The deposits lie in a remarkably straight and narrow zone trending easterly on the south limb of a major isoclinal syncline. The strata and the veins strike east-westerly and dip almost vertically. Other details of this area have been described in the literature (See, e. g., Gunning, 1942; and later references).

Canadian Deposits.

Canada is one of the large gold producers, with about 2/3 of the gold coming from Ontario and the remainder more-or-less equally from British Columbia, Northwest Territories, and Quebec. There are roughly 150 gold mines.

The Canadian Shield is the source of most of Canada's gold, primarily in a belt stretching from western Quebec to the Northwest

Territories. Most of the lodes yield only gold, with some minor silver. There is very little placer gold, probably because there has been practically no seismotectonomagmatic-belt activity in this region during the Proterozoic and Phanerozoic.

Among the primary Canadian deposits are those at the well-known Porcupine, Kirkland Lake, and Little Long Lac districts. Many other gold districts occur within the Precambrian shield, primarily within areas of Keewatin volcanics cut by silicic intrusives. These include Matachewan-Larder Lake, Red Lake-Patricia, Rouyn-Chibougamou, Southeast Manitoba, and Yellowknife. Details of these and other Canadian deposits have been repeatedly discussed in the literature (See, e. g., Bateman, 1950; Hester, 1972). Representative deposits of these districts are discussed in other parts of this chapter.

Central Aldan (USSR).

See Chapter 7.

Central City (Colorado).

See Chapter 10.

Colfax County (New Mexico).

Colfax County, just south of the Colorado border, has been an important source of lode and placer gold. Most of the lode gold of this area has come from the Baldy district. The principal orebodies are in pockets and stringers within basal conglomerates along an unconformity. Minute fissures, filled with ore minerals, interlace the shale and extend upward into the conglomerate. The ore minerals are chalcopyrite, galena, pyrite, and sphalerite. Native gold is present as "wire gold" and oddly-shaped masses. Most of the deposits are associated with the porphyry intrusive bodies. See also New Mexico.

Colorado (USA).

See Chapter 6.

Coolgardie (Australia).

See Western Australia.

Dahlonega (Georgia).

This is part of the Appalachian complex (q. v.).

Deadwood/TwoBit (South Dakota).

Both placers and lodes have been productive in this district that includes mines along the Deadwood Two-Bit, Strawberry, and Elk creeks. Most of the gold comes from placers in Deadwood Gulch. These comprise auriferous conglomerates overlying the Precambrian rocks. Pebbles and small boulders of quartz, quartzite, and schist ore cemented by iron sulfide or oxide where the conglomerate is gold bearing. Some gold comes also from replacement deposits in dolomite beds. Some of the deposits near Strawberry Gulch occur in the Tertiary eruptives and in the adjacent brecciated Precambrian and Cambrian rocks. Most of the ore is auriferous limonite fissure fillings grading downward into galena, pyrite, and copper sulfide.

Delamar (Nevada).

See Chapter 6.

Detour Lake (Ontario).

See Chapter 10.

Dikdik (Ontario).

The Dikdik mine occurs at the contact of granite and granodiorite with lavas. The vein is partly in the intrusive, partly in the lavas. Although small, the veins contain unusually large amounts of gold (Bruce, 1942).

Dome (Ontario).

See Chapter 10.

Flåt (Norway).

The gold deposits at Flåt generally resemble the other gold deposits of Scandinavia (Schønwordt, 1974).

Franklin (Georgia).

This is part of the Appalachian complex (q. v.).

Garden/Maitland (South Dakota).

The Garden district is in the southeastern part of the Precambrian core of the Black Hills deom. Bedrock comprises slate, schist, and quartzite of Precambrian and Cambrian age, plus Tertiary intrusive quartz monzonite porphyry and rhyolite. The orebodies are replacement bodies in dolomite beds in the lower part of the Cambrian Deadwood formation, just above the basal conglomerate. Fractures, parallel to the foliation of the Precambrian rocks, extend upward into the Cambrian Deadwood formation. These appear to have provided paths of preferential accent for the mineralizing solutions. The primary ore comprises finely-divided gold and silver in pyrite, but most of the mined gold is oxidized auriferous and argentiferous limonite.

Gebeit (Sudan).

See Chapter 8.

Georgia (USA).

The gold deposits of Georgia lie within several northeasterly-trending mineralized zones in the crystalline schists and gneisses that underlie the northern third of the State. Many granitic bodies have intruded the foliated rocks. The gold is associated with some dike-type intrusions. Among the many gold deposits within the mineral belt, only three are important: Creighton (in Cherokee County), Dahlonega (Lumpkin), and Nacochee (White). Production is from both lodes and placers. Other details are contained in the literature (See, e. g., Park, 1942; Koschmann and Bergendahl, 1968).

According to the seismotectonomagmatic-belt concept, the Georgia mineral belt is a vestige of "original" Archean gold that was "reworked" during seismotectonomagmatic-belt activity that occurred in Proterozoic and Phanerozoic times.

Ghana (Africa).

These quartz-lode and vein-zone deposits are associated with deep faults along the boundaries of synclinal depressions containing auriferous conglomerates. The mineralization is related to minor intrusions of granitoid rocks intruded along deep faults.

Gold Hill (North Carolina).

See Rowan County.

Golden Mile (Western Australia).

See Kalgoorlie.

Great Fingall (Western Australia).

See Murchison.

Gwalia (Western Australia).

See Leonora Center.

Hahns Peak (Colorado).

See Chapter 6.

Haile (USA).

This is part of the Appalachian complex (q. v.).

Hard Rock (Ontario).

This resembles MacLeod-Cockshutt (q. v.).

Hartley (Rhodesia).

See Chapter 10.

Hill City (South Dakota).

The Hill City district contains widely scattered gold deposits near the headwaters of Spring Creek. The district lies along the west side of a mass of the Precambrian rocks that form the core of the Black Hills. The oldest rocks comprise complexly folded and distorted schist and quartzite. These are intruded by the Precambrian Harney Peak granite. The lode deposits occur in quartz fissure veins and lenses of mineralized shear zones. The veins are mainly quartz with irregularly interspersed gree gold. Some veins contain masses and streaks of lead-antimony sulfide, arsenopyrite, pyrite, some sphalerite, and free gold. The mineralized shear zones are mainly within a northwesterly-trending belt.

Hog Mountain (Alabama).

This is part of the Appalachian complex (q. v.) of Alabama (q. v.).

Homestake (South Dakota).

The ore shoots forming this deposit are confined mainly to the Homestake Formation, a metamorphosed iron formation of Precambrian age. The dominant rock type, cummingtonite schist, is interspersed with some cherty quartz. The orebodies are groups of pipelike southeasterly-plunging shoots, some of which coalesce with other shoots. They are localized in zones of cross folds superimposed on the large regional isoclinal folds (Noble, 1950; Slaughter, 1968). Ores appear to have become concentrated from paths of preferential flow that developed within an active seismotectonomagmatic belt (Tatsch, 1972a). Age data are conflicting, but a Precambrian age is usually assumed on the basis of the deep-seated nature of the mineralization. A Tertiary age is assigned by some researchers on the basis of a small amount of ore that cuts Tertiary igneous rocks, but stable-isotope data (Rye and Rye, 1974) suggest a Proterozoic metamorphic episode about 1.7 b. y. ago. According to the seismotectonomagmatic-belt concept (Chapter 4), the "original" formation probably was associated with a seismotectonomagmatic belt that was active in that area during the Archean. Subsequent metamorphic episodes that were associated with active seismotectonomagmatic belts, primarily during the Proterozoic and Tertiary, served to redeposit some of the original Archean gold.

Further details regarding the Homestake mine are included in the literature (See, e. g., Allsman, 1940; Bayley, 1970; Noble, 1950; Rye and Rye, 1974; Rye et al., 1974; Stemprok, 1968).

<u>Jacobina</u>

See Serra de Jacobina.

<u>Kalgoorlie</u> (Western Australia).

The Golden Mile once made this area the richest gold camp of the world. The deposits are mainly gold-pyrite-telluride replacement lodes, primarily in a regionally-folded Precambrian greenstone belt that contains auriferous quartz dolerite sills. Cross-folding appears to have helped localize the controlling fractures. The fracture intersections contain most of the richest pipes. The Eastern and Western lode systems comprise the Golden Mile. The Eastern system comprises swarms of smaller veins and pipes with local pockets of rich deposits within an essentially irregular distribution pattern. The Western system comprises a few large lodes with regular orebodies and associated gold.

The Oroya shoot is a flat pipe, almost 2 km long, localized by two nearly flat shears within a minor fold of the main Kalgoorlie syncline. The gold is within the structurally controlled ore shoots. The shoots bottom into the underlying, barren schists at a depth of about 1 km. The ore minerals include yellow sulfides, native gold, and tellurides. Associated minerals are galena, sphalerite, enargite, pyrargyrite, magnetite, fluorspar, tourmaline, carbonates, and minor amounts of others.

Other details are contained in the literature (See, e. g., Oversby, 1975; Lamey, 1966; and Tomich, 1974). See also Chapter 10.

Kenterlau (USSR).

The Kenterlau massif is part of the Balkhash anticlinorium in Kazakhstan. The fold is associated with the listwanitization that has occurred within the upper Proterozoic harzburgite (Gavrikova and Larionov, 1970). Besides gold, other mineralizations include cobalt, mercury, and nickel. The harzburgite alteration was produced by postmagmatic solutions accompanying the intrusion of granite porphyry during an episode of seismotectonomagmatic-belt activity in that area. The alteration produced the zoning: serpentinite-talcite-listwanite. The listwanite, stable within a wide PT range, is characterized by the paragenesis: quartz-calcite-dolomite.

Kerr-Addison (Canada).

See Chapter 10.

Keystone (South Dakota).

The Keystone district lies on the northeastern side of the Precambrian Harney Peak batholith near the east margin of the older

Black Hills core. The Precambrian rocks comprise schist, quartzite, amphibolite, plus many granite and pegmatite dikes. The rocks were tightly folded and cut by faults and shear zones during one or more episodes of seismotectonomagmatic-belt activity. Some of the faults and shear zones are mineralized.

The orebodies are in quartz veins or lenticular replacement deposits trending parallel to the foliation of the enclosing schist. One long vein contains white quartz gangue with sparse flake gold. The main minerals are arsenopyrite, pyrrhotite, pyrite, chalcopyrite, sphalerite, and native gold. The ore deposits appear to be Precambrian in age and related to the Harney Peak Granite (Koschmann and Bergendahl, 1968).

Kilo (Zaire).

See Chapter 10.

Kirkland Lake (Quebec).

See Chapter 10.

Korea (Asia).

See Chapter 7.

Lead (South Dakota).

The Lead district contains the famous Homestake mine (q. v.), the leading gold producer in the United States. Rocks of the district are mainly Precambrian, comprising 6 formations aggregating a thickness of 6 km. Some remnants of the Cambrian Deadwood cap some of the hills and ridges. Igneous rocks are of two ages: (1) Precambrian gabbroic rocks, and (2) small dikes, plugs, and stocks of Tertiary porphyries ranging from granite to syenite. The Precambrian rocks are mainly iron-magnesium schistose (lower) and argillaceous phyllites and schists (upper). They are isoclinally folded as a result of Precambrian seismotectonomagmatic-belt activity and they were further deformed during intrusions associated with seismotectonomagmatic-belt activity of the Tertiary.

The orebodies are linear or pipelike replacements in the Precambrian Homestake formation. Four distinct stages of mineralization are discernible. Gold is associated with arsenopyrite, with chloritized rock along the walls of quartz veins, and with the chlorite-pyrrhotite schist that surrounds conformable quartz masses. The orebodies are localized in zones of younger cross folds superimposed on the older isoclinal folds (See, e. g., Noble, 1950).

The mineralization within the Homestake deposits appears to have been produced by at least two episodes of seismotectonomagmatic-belt activity. The earliest of these was during the Precambrian; the latest, during the Tertiary. See further discussion of the applicable seismotectonomagmatic-belt activity under the heading, South Dakota.

Leitch (Ontario).

The Leitch gold mine lies about 5 km east of Lake Nipigon, in the Thunder Bay district. The rocks are mostly graywacke that contain lenses of conglometate and iron formation. The beds are closely folded with many drag folds in the less-competent areas. The major structure is a syncline with axial plane overturned to the north and steep plunge to the west (See, e. g., Bruce, 1942). The main vein is in slightly schistose graywacke. The filling is fractured quartz. The fractures run along strike and others run across the vein. The mineralization occurs mainly along strike and comprises pyrite, tetrahedrite, sphalerite, and gold.

Lena-Vitim (USSR).

These Siberian gold ores were formed in connection with the geosynclinal granitoids within miogeosynclinal depressions filled with monotonous sandy-schistose strata of terrigenous deposits. The deposits consist of rich placers apparently derived from the auriferous pyritic schists (Lamey, 1966).

Leonora Center (Western Australia).

The Leonora district contains the Sons of Gwalia mine, a replacement lode in schistose dolerite. The minerals are quartz, gold, and base-metal sulfides. For a general description, see Western Australia.

Little Long Lac (Ontario).

This district, just north of Lake Superior, consists of Keewatin volcanics overlain unconformably by a sedimentary series. The rocks associated with Little Long Lac appear to be Timiskaming sediments. Some small masses of diorite and narrow dikes of diabase cut the sediments. The veins lie within massive arkose or feldspathic quartzite. The main minerals are quartz, plagioclase, sericite, carbonates, chlorite, and kaolin. Graywacke underlies the arkose. Bands of iron formation underlie the graywacke. The sediments lie in a large syncline. Near the mine, the structure is complicated by a large drag fold, trending N85°E, some fractures, and other surficial features such as would be caused by seismotectonomagmatic-belt activity. The ore bodies are lodes of quartz veins, ranging from mm to dm in size. The primary minerals are quartz and gold, plus minor amounts of base-metal sulfides, stibnite, tetrahedrite, bournonite, and berthierite. These and other auriferous deposits of the Canadian Shield, according to the seismotectonomagmatic-belt concept, are genetically related to each other. See also Kirkland Lake and Porcupine districts. Other details have been discussed in the literature (See, e. g., Bruce, 1942; Bateman, 1951; and later references).

MacLeod-Cockshutt (Ontario).

The MacLeod-Cockshutt mines lie about 2 km south of Little Long Lac (q. v.), and only a short distance from Hard Rock on the east. These mines and Bankfield resemble each other and undoubtedly owe their origin, evolution, and present characteristics to the same episode of seismotectonomagmatic-belt activity (Chapter 4).

Malarctic (Quebec).

See Chapter 10.

Manhattan (Nevada).

See Chapter 8.

Maslovskiy (USSR).

See Chapter 8.

Matachewan (Ontario).

Matachewan is about 50 km southwest of Kirkland Lake. The orebodies lie (1) in the Keewatin basalt series near the top in a zone of abundant local flow-tops, (2) in the Keewatin agglomerate schist, especially where it is contorted, and (3) in the Algoman syenite and syenite porphyry (Derry, 1942a). The ore comprises silicified, carbonatized, and mineralized basalt, with fractures filled with quartz and some albite. Within the orebodies and extending beyond some of them are tension fractures, dipping flatly north, filled with quartz, albite, and some pyrite.

McIntyre (Ontario).

See Chapter 10.

McWatters (Quebec).

The McWatters gold deposit, within the Rouyn-Harricanaw belt, about 8 km southeast of Noranda, occurs in a series of Keewatin conglomerate sediments. The sediments are intruded by tuffs, agglomerate, and a few amygdaloidal lava flows. These have been overturned and dip steeply to the north. The mineralization comprises quartz, tourmaline, some (10%) base-metal sulfides, and native gold. The gold appears to have been introduced during the late stages of mineralization, following an episode of seismotectonomagmatic-belt activity in that area at that time (See, e. g., Hawley, 1942).

Meekatharra (Western Australia).

See Western Australia.

Minas Gerais (Brazil).

See Morro Velho.

Miracle Mile (Western Australia).

See Kalgoorlie.

Morro Velho (Brazil).

Morro Velho lies in Nova Lima, Minas Gerais. Production from this mine, begun as an open pit about 200 years ago, developed into underground operations to depths of 2 1/2 km. The deposits are hydrothermal replacements of bedded quart-ankerite-dolomite rock. The gold is associated with sulfides (See, e. g., Gair, 1962). See also Chapter 10.

Moto (Zaire).

See Chapter 10.

Murchison (Western Australia).

These quartz lodes, localized at the Wiluna and Great Fingall mines, are similar to the Kalgoorlie lodes (q. v.).

New Mexico (USA).

The gold-producing districts of New Mexico are distributed along a northeasterly-trending mineral belt that exyends from Hidalgo County in the southwestern corner to Colfax County in the north-central part of the State. This belt is in the mountains lying between the Colorado Plateau on the northwest and the Great Plains on the east. It is a zone of crustal disturbance such as would be expected from seismotectonomagmatic-belt activity. During this activity, the rocks were folded, faulted, and intruded by dikes, laccoliths, and stocks of monzonite rocks. This belt contains deposits of copper, gold, lead, silver, and zinc. Some of the copper and gold deposits date back to the Precambrian. However, most deposits are associated with late Cretaceous and Tertiary intrusives (See, e. g., Koschmann and Bergendahl, 1968). The gold placers appear to have been derived from the weathering of these deposits. During the late Tertiary, lavas poured out over the district, and fissures within these rocks became mineralized. The fissures are rich in gold and silver but they are relatively poor in base metals.

The most productive of New Mexico's districts are in the counties of Catron, Colfax, Grant, and Sierra. These are discussed in other sections of this book, under these county names.

When the northeasterly-trending mineral belt of New Mexico is closely analyzed, it is found that the gold deposits there, like those in other parts of the world, appear to have developed in a manner consistent with the seismotectonomagmatic-belt concept. That is, some of the "original" Archean gold of the Earth's tectonosphere was reworked during the seismotectonomagmatic-belt activity that occurred in Proterozoic and Phanerozoic times. Most of the evidence of the Proterozoic and Phanerozoic activity lies hidden or has been obliterated by subsequent Phanerozoic seismotectonomagmatic-belt activity. However, many vestiges of this activity still lie within this northessterly-trending mineral belt of New Mexico.

Norilsk (USSR).

These complex intrusions of basic magmas with accompanying magmatic and post-magmatic deposits contain auriferous Cu-Ni-Pt ores. Gold occurs in the Norilsk deposits within paragenetic associations in solid solutions of palladium and tin in platinum; within sold solutions of platinum, tin, and lead in palladium and lead, arsenides of pallidium and lead, ferroplatinum, and sperry-stibiopalladinite. Gold mineralization in these associations is younger than is the associated silver-cuproaurite (Razin and Yukina, 1971).

Northern Empire (Ontario).

The Northern Empire area comprises basic lavas, dikes, and irregular diabase masses. The diabase cuts the lavas, which are Keewatin. The main vein has been traced for over 600 m. Some faults offset the vein, usually less than a meter. The vein-filling comprises quartz, with some arsenopyrite, pyrite, and gold. Other details have been discussed (See, e. g., Bruce, 1942; and later references).

Nova Lima (Brazil).

See Morro Velho.

Oroya (Western Australia).

See Kalgoorlie.

Peak Hill (Western Australia).

See Western Australia.

Porcupine (Ontario).

The Precambrian gold-quartz lodes at Dome, Hollinger, and McIntyre resemble those at Ashanti (q. v.) in Ghana.

Rhodesia (Africa).

See Chapter 10.

Rouyn-Harricanaw (Quebec).

The Rouyn-Harricanaw gold belt extends eastward roughly 200 km from the Ontario boundary. The area is structurally complex. The geological succession comprises (1) Keewatin andesitic and rhyolitic lavas, tuffs, and conglomerates, (2) Timiskaming conglomerates and graywackes overlying the Keewatin, (3) earlier quartz diorite intrusives, and (4) later poorly-defined intrusives. The gold ores post-date the intrusives but appear to antedate Keweenawan quartz and olivine diabase dikes. The Keewatin and Timiskaming series have been steeply folded along easterly-trending axes. The late diabase dikes cut across these on a northeasterly strike. Complex faulting has occurred along strike and oblique thereto. Except for gold associated with the copper sulfides at Noranda, the gold deposition appears to be pre-diabase.

Structurally, the gold mineralization fills fractures of the shear and tension types. Free gold appears to have been introduced into earlier-formed vein material along fractures, crush zones, and schist planes produced by one or more episodes of seismotectonomagmatic-belt activity in that area. Other details are available in the literature (See, e. g., Hawley, 1942; and later references).

Serra de Jacobina (Brazil).

The metamorphased auriferous conglomerates at Serra de Jacobina, Bahia, produce about 0.5 oz/tn of gold plus some uraninite and platinum-group metals. Also known as fossil placers, these deposits are vestiges of "original" Archean gold that was "reworked" by seismotectonomagmatic-belt activity during the Proterozoic (Tatsch, 1973a). The conglomerates have become lithified and welded to the bedrock. Comprising quartz pebbles imbedded in pyrite and micaceous minerals, these conglomerates resemble fossil placers in other parts of the world, including Tarkwa (q. v.) and Witwatersrand (q. v.). Other details are contained in the literature (See, e. g., Stemprok, 1968; Brobst and Pratt, 1973).

<u>Siscoe</u> (Quebec).

The Siscoe gold deposit is on a small island of De Montigney Lake, about 85 km east of Noranda. The deposit occurs in an intricate but well-defined fracture system. The mine is mostly in a small highly-altered Precambrian stock, comprising an early (basic) and later phase of albite and quartz. This stock intrudes the steeply-dipping Keewatin lavas of intermediate-to-basic composition and extending easterly in a 30-km-wide belt. Other granodiorite intrusions are in the area. Gold-quartz-tourmaline veins are within and adjacent to the Siscoe and Bourlamaque intrusives. Fractures are tensional and shear. The fracture fillings comprise: (1) quartz with minor base-metal sulfides, (2) later dikes of syenite porphyry, albite, andesite, and fine diorite, and (3) quartz, tourmaline, and the bulk of the gold mineralization. The latter penetrates the veins, albite, and fractures.

As is the case with other mines of that area, the geology and mineralization of Siscoe are derivable from one or more episodes of seismotectonomagmatic-belt activity in that area, dating from the Archean.

<u>South Dakota</u> (USA).

The gold deposits of South Dakota form a short, compact northwesterly-trending belt in the Black Hills within Lawrence and Pennington counties of the western part of the State. The largest producer is Homestake. Others are the Bald Mountain, Deadwood-Two-Bit, Garden, Hill City, Keystone, and Squaw Creek districts. These are described within separate entries under the indicated county names.

The Black Hills comprise a core of Precambrian rocks surrounded by concentric bands of Paleozoic and Mesozoic rocks that dip gently away from the core. They are intruded by igneous dikes, sills, and stocks ranging from rhyolite and phonolite to quartz monzonite. The Precambrian rocks are exposed in a northwesterly-trending oval area about 100 km long and 40 km side. These rocks comprise (1) intensely folded metasedimentary rocks such as schist, slate, and quartzite, (2) amphibolites cutting the metasediments, and (3) granite intruding (1) and (2). The famous Homestake orebody occurs within a cummingtonite schist that lies stratigraphically about 600 m above the base of the Precambrian rocks exposed in that area. The main period of deformation that produced the Black Hills dome occurred during the post-Cretaceous Laramide orogeny. The structure appears to have resulted from vertical forces, such as those that would be inherent in seismotectonomagmatic-belt activity.

The Black Hills gold deposits fall into 5 general classes: (1) replacement deposits in Precambrian rocks; (2) quasi-placers in Cambrian basal conglomerates; (3) Tertiary replacements in Cambrian and Mississippian rocks; (4) deposits in Tertiary eruptives; and (5) Recent placers. A closer scrutiny of these five classes of deposits reveals that the South Dakota gold deposits, like those of other parts of the world, have developed in a manner consistent with the predictions of the seismotectonomagmatic-belt concept. That is, some of the "original" Archean gold of the Earth's tectonosphere was "reworked" during seismotectonomagmatic-belt activity that occurred in Proterozoic and Phanerozoic times.

Staryy Oskol (USSR).

The Staryy Oskol gold deposits, in south central USSR, lie within the Voronezh anticline comprising Precambrian basement rocks (D'Yachenko et al., 1970). Most of this gold is finely divided (4 to 6 microns) within the sulfides that comprise 30% of the core rocks of the Tim-Yastrebovka synclinorium associated with the Staryy Oskol deposits. The sulfides are mainly pyrite plus some chalcopyrite, sphalerite, marcasite, pyrrhotite, galena, and arsenopyrite.

Sturgeon River (Ontario).

The gold mine at Sturgeon River resembles the Dikdik mine (q. v.).

Sudbury (Canada).

These complex intrusions of basic magmas with accompanying magmatic and post-magmatic deposits contain auriferous Cu-Ni-Pt ores.

Taber (Virginia).

This is part of the Appalachian complex (q. v.).

Tarkwa (Ghana).

The metamorphosed auriferous conglomerates at Tarkwa occur in terrigenous deposits forming a narrow graben-cyncline comprising intensely-faulted rocks. Also known as fossil placers, these deposits formed in the Proterozoic and have been lithified to conglomerate to become part of the bedrock. These conglomerates comprise quartz pebbles embedded in pyrite and micaceous minerals. Besides gold, they contain also uraninite and the platinum group metals. The gold yield is about 0.2 oz/tn. Other details are contained in the literature (See, e. g., Stemprok, 1968; Brobst and Pratt, 1973). The world's best known fossil placers are in the Witwatersrand complex (q. v.).

Tennessee (USA).

Almost all of Tennessee's gold comes from placer deposits in Monroe County and from the Ducktown copper ores of Polk County, both in the southeastern corner of the State. The Monroe county placers are in high terrace gravels, stream gravels, and alluvial deposits. The country rocks in the Ducktown district comprise graywacke, conglomerate, mica schist, slate, staurolite schist, and garnet schist. All these are part of the Precambrian Great Smoky group. They are cut be several gabbro dikes. The rocks have been intricately folded, but the main pattern involves 3 northeasterly-plunging anticlines carrying many drag folds. Three fault systems, some preore and some postore, cut the rocks.

The ore deposits are tabular or lens-shaped replacement bodies in mica schist that have been complexly folded and faulted from seismotectonomagmatic-belt activity in that area. Some observers consider the deposits to be multistage hydrothermal vein deposits. There are 3 types of ore: (1) surficial gossan rich in hydrous iron oxides, (2) secondary sulfide ores just below the water table, and (3) the primary sulfide ore. Small amounts of gold and silver occur in the primary and secondary sulfide ores.

Thunder Bay (Ontario).

The Tunder Bay district extends from Lake Superior northward to the Albany River area. The mines of the district include Bankfield (q. v.), Dikdik (q. v.), Hard Rock (q. v.), Leitch (q. v.), Little Long Lac (q. v.), MacLeod-Cockshutt (q. v.), Northern Empire (q. v.), and Sturgeon River (q. v.). Other details of the district have been described in the literature (See, e. g., Newhouse, 1942; and later references).

Timmins (Ontario).

See Chapter 10.

USSR Placers.

See Chapter 8.

Val d'Or (Quebec).

See Chapter 10.

Western Australia.

The Western Australia gold is primarily from lodes. The area is underlain by Precambrian metamorphic and sedimentary rocks intruded by gneissic granites. Most of the deposits lie in minor pockets of invaded rocks. The deposits are of 4 main types: (1) fissure lodes, (2) auriferous sediments, (3) quartz reefs, and (4) auriferous dikes.

The long, deep fissure veins are primarily replacements along shear zones in basic metamorphic rocks, such as would be expected to have developed within an active seismotectonomagmatic belt. The minerals are pyrite and base-metal sulfides, with some minor tellurides, scheelite, vanadinite, and bismuthinite. The auriferous sediments are replacements by silica with much low-grade auriferous pyrite. The quartz reefs are essentially fissure fillings of quartz and base-metal sulfides. The gold is localized in short, shallow, rich ore sheets. The auriferous dikes are primarily aplites and alaskites. These dikes consist primarily of low-grade auriferous quartz, with some iron sulfides, molybdenite, wulfenite, and crocoite in the alaskites; and of silicates, pyrrhotite, and gold in the aplites.

Some of the well-known deposits of Western Australia include the Golden Mile of Kalgoorlie (q. v.), Sons of Gwalia mine in the Leonora Center (q. v.), the Wiluna and Great Fingall lodes of the Murchison field (q. v.). Others are the Meekatharra lodes, the Yilgam quartz hematite schists and auriferous sediments, and the sediments of Peak Hill, Coolgardie, Big Bell, and Cue.

Wiluna (Western Australia).

See Murchison.

Witwatersrand (Africa).

These metamorphosed auriferous conglomerates are confined to thick strata of terrigenous deposits that fill a large syncline. The structure is complex due to fractures and dome-shaped prominences of the basement. The gold occurs in the beds of the Witwatersrand system, which occupy a syncline extending about 175 km in a generally east-west direction. The Witwatersrand system, about 8 km thick, comprises a lower and upper division. The lower Rand (4.5 km) comprises alternating zones of shale (or slate), quartzite and minor conglomerates. The upper Rand (3.5 km) comprises quartzite, "reef" conglomerates, and shale at the base.

As is the case with most other gold deposits of the world, the origin of the Witwatersrand gold deposits is not understood. Two major concepts have been advanced in the literature; an alternate concept is provided by the seismotectonomagmatic-belt concept (See, e. g., Tatsch, 1973a). The two concepts from the literature (See, e. g., Borchers, 1961; Davidson, 1960; Lamey, 1966; Whiteside and Brock, 1961) may be summarized:

a. The syngenetic concept. In this concept, the gold, uranium, and much of the pyrite originally accumulated together with the pebbles and matrix of the conglomerate. Subsequent modification resulted from low-grade metamorphism, metasomatism, an other short-transport rearrangements.

b. The epigenetic concept. In this concept, the gold, uranium, and much of the pyrite and other sulfides were introduced by hydrothermal solutions of magmatic origin, subsequent to the deposition of the conglomerate.

According to the seismotectonomagmatic belt concept, these two concepts are subsets of a larger system, which may be expressed schematically:

$$\sum_{i=1}^{n} a_i A_i \quad n=3, \text{ where}$$

A_1 = syngenetic deposition

A_2 = epigenetic deposition

A_3 = seismotectonomagmatic modification subsequent to A_1

a_i = parameters, any of which may be zero, provided that

$$\sum_{i=1}^{n} a_i = 1.$$

The seismotectonomagmatic-belt concept is discussed more fully elsewhere (See, e. g., Tatsch, 1973a). Other details of the Witwatersrand deposits are available in summary discussions (See, e. g., Bateman, 1951; Brobst and Pratt, 1973; Lamey, 1966; Stemprok, 1968).

Wyoming (USA).

The two main gold-bearing districts of Wyoming are Atlantic-City / South-Pass and Douglas Creek. Gold placers occur along Douglas Creek and its tributaries. Lode deposits occur within Precambrian schist veins. These veins, containing chalcopyrite, gold, and pyrrhotite, were worked for copper, silver, and gold.

The Atlantic-City / South-Pass district has produced most of Wyoming's gold from both placers and lodes. The bedrock comprises mainly isoclinally folded and faulted Precambrian amphibolite, diorite, gneisses, granodiorite, and granite. Auriferous veins have been found in fault zones in all types of country rock. These veins occur in an en-echelon pattern in zones extending over 300 m. The unoxidized ore minerals are arsenopyrite, pyrite, and some scattered chalcopyrite and galena.

Yenisei (USSR).

These gold ores of Siberia were formed in connection with the geosynclinal granitoids within miogeosynclinal depressions filled with monotonous sandy-schistose strata of terrigenous deposits. The gold-bearing quartz veins are similar to those at Ballarat (q. v.) and Bendigo (q. v.) of Australia (Lamey, 1966).

X-ray analyses of the different types of arsenopyrites in the Yenisei ridge, like those in many other parts of the world, show that the type formed depends on the temperature. Thus, the arsenopyrites near the associated granitic intrusion are of the high-temperature type, while those farther away are of decreasingly lower-temperature types. Unlike the deposits near the intrusion, those farther away contain a low-temperature polymetallic association (See, e. g., Li et al., 1971). These observations are consistent with the predictions of the Tectonospheric Earth Model (Tatsch, 1972a, 1973a).

Yilgam (Western Australia).

See Western Australia.

Zaire (Africa).

See Chapter 10.

Conclusions Regarding the Proterozoic Gold Deposits.

Close scrutiny of the known Proterozoic gold deposits reveals that almost all of them appear to be associated with those areas of the Earth's surface where vestiges of Proterozoic seismotectonomagmatic-belt activity remain today. These areas are found primarily within the platforms and their folded peripheries, including those of western Siberia, southern and western Africa, the United States, Canada, Brazil, India, Scandinavia, and Australia.

In comparing the above-listed Proterozoic gold deposits with the predictions of the Tectonospheric Earth Model, it is well to recall the comments that were made in the first section of this chapter. Those comments, taken in conjunction with the concepts of Chapters 2, 3, 4, and 5, should permit the reader to associate each Proterozoic gold deposit with the corresponding vestiges of Proterozoic seismotectonomagmatic-belt activity. In this oonnection, the general comments at the end of Chapter 6 apply also here. This includes the invitation that the reader correspond directly with the author in those cases where either the concepts or their application to specific Proterozoic gold deposits is not clear to the reader.

Chapter 10

ARCHEAN SEISMOTECTONOMAGMATIC BELTS AND THE ASSOCIATED GOLD DEPOSITS

In considering the Archean gold deposits, it is well to recall the basic enigmas that exist within our knowledge of the Archean. These include: (1) the remarkable similarities between the South African, Canadian, and Australian Archean shields (See, e. g., Oversby, 1975); (2) the differences between the Archean shields and younger rocks; (3) the evidence that the geological processes operating during the Archean were not exactly the same as they are today (See, e. g., Anhaeusser et al., 1969); (4) the large volumes of acidic plutonic and migmatitic rocks that were produced at a time when the Earth is usually assumed to have had only a thin, primitive crust; and (5) the fact that the source of material for the Archean rocks, particularly the more potassic ones, is not obvious when considered in connection with traditional Earth models. These and other Archean enigmas contributed, some years ago, to the author's realization that the early history of the Earth can be understood only in terms of a deep-seated, global driving mechanism that has been operating during the 4.6 b. y. that the Earth is believed to have been in existence (See, e. g., Tatsch, 1963b).

The gold deposits associated with the Archean seismotectonomagmatic belts comprise, according to the Tectonospheric Earth Model concept, part of the Earth's "original", early-Archean gold that was reworked by subsequent seismotectonomagmatic-belt activity during the Archean but not during the Proterozoic and Phanerozoic. It is not too surprising that the Earth's Archean gold deposits are found primarily within the shields of Africa, Asia, Australia, and America (Sistrini, 1973). These are the main areas where vestiges of Archean seismotectonomagmatic-belt activity still remain today. These vestiges occur mainly as long, rectilinear belts crisscrossing the Archean cratons (See, e. g., Oversby, 1975).

In considering the distribution patterns of the Earth's Archean gold deposits, it is well to bear in mind that the late-Archean shield areas were not necessarily the same as they are today, nor the same as they were during the early Archean. According to the Tectonospheric Earth Model, the words of Heraclitus, "panta rhei" (everything changes), represent the dynamic behavior of the Earth during the past 4.6 b. y., and particularly so during the Archean (Tatsch, 1972a).

Thus, the Canadian shield could have been separated into 2 or more parts during the late Archean and early Proterozoic. For example, an archeoplate comprising the Churchill, Superior, and Southern provinces was probably separate from an archeoplate comprising the Bear and Slave provinces (See, e. g., Symons, 1975). Likewise, the archeoplates in other parts of the world must be analyzed in a similar manner before analyzing the origin and evolution of today's gold deposits.

Eurasia provides an excellent example. Today, Eurasia comprises a multiplicity of cratonic units of various ages (See, e. g., Tatsch, 1972a: chapter 7). These cratons include (1) the north European plate, (2) the south European plate, (3) the Siberian platform, (4) the Jano-Kolymian plate, (5) the Kazakhstan plate, (6) the north Chinese plate, and (7) the south Chinese plate. Separating these cratonic units from each other, and traversing at least some of them intracratonically, are fossil orogens, sutures, and other vestiges of erstwhile seismotectonomagmatic-belt activity dating back to the Archean. These 7 plates could easily have been parts of separate continental entities at least once during the Archean and perhaps many times during the Proterozoic and Phanerozoic (Tatsch, 1972a: chapter 7).

Representative Archean Gold Deposits.

In order to determine how well today's actual gold deposits follow the patterns predicted by the Tectonospheric Earth Model, it is well to consider the locations and salient characteristics of representative examples of known Archean gold deposits.

Aldan Shield (USSR).

See Chapter 7.

Anabar-Olenek (USSR).

The gold deposits of Anabar-Olenek are associated with the platinum-group minerals in the placers within the basins of Buolkaloah, Khatygyn-Yelete, Arakh-Bilire, and other streams in the lacustrine and alluvial sediments of Quaternary age (Shpunt, 1970). The source of the placers appears to be the Proterozoic block-faulted terrigenous series cropping out near the deposits. The Proterozoic rocks contain detrital chromite, spinel, and pyroxene. Closer scrutiny suggests that the Anabar-Olenek deposits were developed according to the seismotectonomagmatic-belt concept. Thus, these deposits include "original" Archean gold that was "reworked" during the Archean and early Proterozoic, as well as since then in some areas.

Ashanti (Ghana).

The Precambrian gold-quartz lodes at Ashanti resemble similar ones in all parts of the world, including Homestake (South Dakota), Central City (Colorado), Porcupine and Kirkland Lake - Larder Lake (Ontario), Morro Velho (Brazil), Kolar (India), Kalgoorlie (Western Australia), Bulawayo and Hartley (Rhodesia), Kilo and Moto (Zaire), and various other deposits in the shields of Siberia, Canada, Brazil-Guyana, Africa, and Arabia. Although there are variations within these deposits, they are essentially hydrothermal veins of quartz and gold (See, e. g., Brobst and Pratt, 1973). In some, like Homestake, the gold and quartz replace the wall rock; in others, the gold and quartz fill open spaces along fracture zones. In at least some cases, the wallrock-replacement type is older. Most of these deposits were formed at great depths and still lie as much as several km beneath the present surface. All of them are enclosed in Precambrian rocks and are of Precambrian age. The enclosing rocks are mainly metamorphosed volcanic and sedimentary rocks of the greenstone-granite association.

The gold content of the ores ranges from about 0.3 to about 0.6 oz/tn. The fact that the auriferous content of these ores falls within such close limits suggests: (1) that they comprised original Archean gold that has been reworked during late Archean and early Proterozoic seismotectonomagmatic-belt activity (Tatsch, 1973a); and (2) that they have been affected by little or no seismotectonomagmatic-belt activity since the Precambrian.

These two suggestions are further substantiated by total production; that is, the <u>oldest</u> of the Precambrian gold-quartz lodes, generally, have produced the <u>most</u> gold, whereas the <u>youngest</u> Precambrian gold-quartz lodes have produced relatively <u>less</u>. Among the oldest, large producers are two that have produced more than 20 million oz (i. e., Homestake and Kolar) and 6 that have produced from 5 to 20 million oz each (i. e., Dome, Hollinger, Kerr-Addison, Kirkland Lake, McIntyre, and Morro Velho). Younger Precambrian gold-quartz lodes, generally, have produced less than 5 million oz each; examples of these are Alaska Juneau, Ballarat, Bendigo, Browhhill-Oroya, Empire, and North Star. Interpreted by the seismotectonomagmatic-belt concept, the oldest Precambrian gold-quartz lodes have produced more gold because the "original" Archean in them has been less "reworked" by late Archean and subsequent seismotectonomagmatic-belt activity than has the gold in the younger Precambrian gold-quartz lodes (Tatsch, 1973a).

Brownhill-Oroya (Australia).

The Precambrian gold-quartz lodes at Brownhill-Oroya, in Kalgoorlie, resemble those at Ashanti (q. v.), in Ghana.

Bulawayo (Rhodesia).

The Precambrian gold-quartz lodes at Bulawayo resemble those at Ashanti (q. v.), in Ghana.

Central Aldan (USSR).

See Chapter 7.

Central City (Colorado).

The Precambrian gold-quartz lodes at Central City resemble those at Ashanti (q. v.), in Ghana.

Champion Lode (India).

See Kolar.

Colorado (USA).

The Precambrian gold-quartz lodes at Central City, Colorado, resemble those at Ashanti (q. v.), in Ghana.

Detour Lake (Ontario).

This new discovery lies just inside Ontario on the Ontario-Quebec line about 200 km north of Kirkland Lake. Initial estimates suggest that the 0.2 oz/tn deposit should yield at least 1 million oz, perhaps 10 times that much (See, e. g., Brown, 1975). The ore falls within an elongate belt at least 2 km long and from a meter to 150 m in width. Rock outcrops are rare in this swampy area. The overburden is from 10 to 40 m thick. The Selco base-metal discovery is nearby, but very little copper or other massive sulfides are present within the auriferous deposit. The gold mineralization is quite fine and mostly in the free state. Closer scrutiny suggests that the cupriferous and auriferous deposits in the Detour Lake area are geologically separate. The gold deposit occurs within mafic volcanics, whereas the nearby base-metal deposits are within more-acid, rhyolitic-type rocks.

Depthwise, the gold deposit at Detour Lake resembles those at Timmins, Kirkland Lake, Noranda, and Malarctic.

Dome (Ontario).

The Precambrian gold-quartz lodes at Dome, in the Porcupine area, resemble those at Ashanti (q. v.), in Ghana.

Flin Flon (Canada).

This base-metal deposit furnishes much of the gold of Manitoba and Saskatchewan as a byproduct. The deposit, a replacement, lies in Proterozoic sheared volcanic rocks. The principal ore bodies are masses of solid sulfides, chiefly pyrites with some chalcopyrite and sphalerite. Outputs include gold, silver, copper, zinc, and some cadmium, tellurium, and silenium. According to the seismotectonomagmatic-belt concept, the Flin Flon gold is part of the "original" Archean gold that was "reworked" during Archean and early Proterozoic seismotectonomagmatic-belt activity in that area.

Granada (Canada).

The Granada gold mine, in Quebec about about 6 km south of Noranda, resembles the Kirkland Lake district mines (q. v.). According to the seismotectonomagmatic-belt concept, these mines are genetically related.

Hartley (Rhodesia).

The Precambrian gold-quartz lodes at Hartley resemble those at Ashanti (q. v.), in Ghana.

Hollinger (Canada).

This is part of the Porcupine district. One vein system in this mine is 2 km long (Lamey, 1966). See also Ashanti.

Kalgoorlie (Australia).

The Precambrian gold-quartz lodes at Browhhill-Oroya, in Kalgoorlie, resemble those at Ashanti (q. v.), in Ghana.

Kerr-Addison (Canada).

The Precambrian gold-quartz lodes at Kerr-Addison, in the Larder Lake area, resemble those at Ashanti (q. v.), in Ghana.

Kilo (Zaire).

The Precambrian gold-quartz lodes at Kilo resemble those at Ashanti (q. v.), in Ghana.

Kirkland Lake (Canada).

Similar to Porcupine, this Quebec district consists of several contiguous producers along a single belt, including Kirkland Lake Gold, Lake Shore, Macassa, Sylvanite, Teck-Hughes, Toburn, and Wright-Hargreaves.

The deposits lie within a synclinal belt of metamorphosed Temiskaming sediments, underlain by Keewatin volcanics, and intruded by syenites and basic dikes. Parallel reverse faults have localized the ore in the above system. Post-ore movement produced diagonal cross and strike faults that displace the veins. At some places, the area between parallel fractures is mineralized to form wide, banded ore bodies or breccia masses. Some shoots are roughly 2 km deep, Native gold and tellurides are associated with pyrite in quartz, calcite, and altered rock. The tellurides are altaite, calaverite, coloradite, and petzite. Some tourmaline and sparse base-metal sulfiees are also present.

The intense silicification of the ore zone and carbonization of the wall rocks and ore minerals suggest a genetic relationship to a deep-seated magma whose differentiation supplied the intrusives and ore solutions (See, e. g., Bateman, 1950). According to the seismotectonomagmatic-belt concept, the deposits of the Kirkland Lake district and of the Porcupine district were produced by the same seismotectonomagmatic belt. Thus they both belong to the same metallogenic epoch and province.

Kolar (India).

The groups of quartz and sulfide lodes associated with this ore field in Mysore are identified with a narrow 80-km-long syncline comprising amphibolites of the Dharwar series. The ore minerals of the Kolar veins are scheelite, magnetite, ilmenite, pyrite, galena, pyrrhotite, arsenopyrite, chalcopyrite, native gold, old telluride, and silver.

The production is derived from a single reef, the Champion Lode. The ore is concentrated in shoots that pitch about 45° within the vein. The fissuring appears to be due to faulting. The localization of the ore shoots appears to be drag-fold controlled. The mineralization solutions appear to have been associated with the magmatic sources that produced the later granite. Other details of the Kolar field are discussed by Viswanathan (1974), Rozhkov (1970), Bateman (1951), Lamey (1966), Stemprok (1968), Brobst and Pratt (1973), and others. See also Ashanti.

Larder Lake (Canada).

This Quebec deposit is similar to Porcupine.

Malarctic (Canada).

Malarctic lies in Quebec about 65 km east of Noranda. The Precambrian rocks include some Temiscamian graywacke sediments that dip steeply northward. These are intruded by a few basic dikes and by younger stocks and dikes of feldspar porphyry. The main gold ore bodies comprise altered and mineralized graywacke. Other details have been discussed in the literature (See, e. g., Derry, 1942b; and later references).

McIntyre (Ontario).

The Precambrian gold-quartz lodes at McIntyre in the Porcupine area resemble those at Ashanti (q. v.), in Ghana.

Morro Velho (Brazil).

The Precambrian gold-quartz lodes at Morro Velho in Minas Gerais resemble those at Ashanti (q. v.), in Ghana.

Moto (Zaire).

The Precambrian gold-quartz lodes at Moto resemble those at Ashanti (q. v.), in Ghana.

Noranda (Canada).

This base-metal mine, known for its copper, has produced much of Canada's gold. The deposits are generally of two types: (1) sulfide-replacement deposits, and (2) gold-bearing veins. The deposits are associated with faults and replace fractured rhyolite and andesite rocks. The presence of pyrrhotite, tourmaline, and arsenopyrite suggests a high-temperature hydrothermal origin, but some tellurides suggest that at least part of these deposits were deposited at a lower temperature (Lamey, 1966).

Porcupine (Ontario).

This deposit forms part of an elongate belt, at least 300 km long, that contains the outstanding Canadian auriferous formations. Throughout this belt, volcanic and sedimentary rocks of the Archean have been complexly folded, sheared, fractured, faulted, and intruded by granitic and some gabbroic masses and dikes. The belt also contains many faults, suggesting surficial manifestations of a long-lived, deep-seated belt of weakness beneath this area. Most of the mines are clustered about these faults, suggesting that both the faults and the mineralizations were produced within a seismotectonomagmatic belt.

The deposits are quartz veins and lodes with some native gold mixed with silver. Some vein-bound scheelite and pyritic sulfides are found. This district includes about 40 mines. Among them are Buffalo-Ankerite, Delnite, Dome, Hallnor, Hollinger, McIntyre, Parmour, Paymaster, and Preston East Dome. Other details are included in the literature (See, e. g., Bateman, 1951; Brobst and Pratt, 1973; Hurst, 1942; and Lamey, 1966). See also Ashanti.

Rhodesia (Africa).

The Precambrian gold-quartz lodes at Bulawayo and Hartley, in Rhodesia, resemble those at Ashanti (q. v.), in Ghana.

Timmins (Ontario).

This deposit is similar to Porcupine (q. v.).

Val d'Or (Quebec).

Located in the Detour Lakes - Strugeon River area, the Val d'Or gold deposits at the new Belmoral Mines show excellent promise (Anon., 1975). This amee belt has produced 50 other mines. The gold averages above 0.17 oz/tn. The new mines lie within a one-km-long shear that widens to the east and has an average width of about 20mm. The mineralization appears to extend to a considerable depth. Intersected ores include chalcopyrite with rhyolites at the extreme north end of the property. This lies adjacent to the contact with the granodiorite batholith. Ninety percent of the mines are found in quartz tourmaline. Many major shears, up to 25 m wide, cut the batholith on an east-west strike.

Yellowknife (Northwest Territories).

The chief producer of this district, in the Mackenzie area, is the Giant Yellowknife mine. Here, early Precambrian volcanic and sedimentary rocks have been complexly deformed, sheared, fractured, faulted, and intruded by gabbroic and granitic rocks. The best gold deposits are in quartz-carbonate lenses and quartz veins localized in schist zones within the greenstone belt. Some lesser deposits occur in metasediments. Gold occurs as native metal with some aurostibnite (= gold antimonide). Quartz is the principal gangue mineral, accompanied by moderate amounts of ankerite and calcite. The metallic minerals include pyrite and arsenopyrite in large, ubiquitously-distributed quantities. Some widespread sphalerite and chalcopyrite deposits comprise only small quantities of these base metals.

As is the case with other deposits, there is no consensus regarding the origin of the Yellowknife deposits. Some researchers consider that the ore and gangue materials were present in country rock and were concentrated into the present deposits by metamorphism and the dilitancy imposed by the great shear-zone system (See, e. g., Boyle, 1959; Boyle et al., 1963; Wanless et al., 1960). Other researchers feel that the deposits were of hydrothermal origin (See, e. g., Coleman, 1957; Ames, 1962). It is interesting that all these investigators used the same data. However, this is not too surprising when these deposits are interpreted in terms of the seismotectonomagmatic-belt concept. In this respect, the origin of the Yellowknife deposits resembles the origin of those at Witwatersrand (q. v.).

Zaire (Africa).

The Precambrian gold-quartz lodes at Kilo and Moto, in Zaire, resemble those at Ashanti (q. v.), in Ghana.

Conclusions Regarding the Archean Gold Deposits.

When the known Archean gold deposits are analyzed, it is found that almost all of them appear to be associated with those areas of the Earth's surface where vestiges of Archean seismotectonomagmatic-belt activity remain today. These are found primarily within the shields of Africa, Asia, Australia, and America.

In comparing the known Archean gold deposits with the predictions of the Tectonospheric Earth Model, it is well to recall the comments made, at the beginning of this chapter, about the Earth's enigmatic behavior during the Archean. Those comments, taken in conjunction with the concepts of Chapters 2, 3, 4, and 5, should permit the reader to associate each Archean gold deposit with the corresponding vestiges of Archean seismotectonomagmatic-belt activity. In this connection, the general comments at the end of Chapter 6 apply also here. This includes the invitation that the reader correspond directly with the author in those cases where either the concepts or their application to the specific gold deposits is not clear to the reader.

Chapter 11

THE PRESENT DISTRIBUTION OF GOLD DEPOSITS AS A FUNCTION OF THE EARTH'S BEHAVIOR DURING THE PAST 4.6 BILLION YEARS

It has been known for some time that not all gold provinces contain the same types of deposits. This observation is probably best explained in terms of regional and global patterns in the periodicity of igneous activity (See, e. g., Ahrens and Schubert, 1975; Anguita and Hernan, 1975: Armstrong, 1975; Cawthron, 1975; Noble, 1974; Tatsch, 1972a, 1973a, 1975a; Taylor, 1975).

Thus, we see, for example, that most porphyry-copper deposits containing significant molybdenum contain relatively small amounts of gold (See, e. g., Kesler, 1973). When an analyses is made on a global basis, over a long period of time, other correlations suggest that crustal type, although important, is definitely not the only controlling factor in the development of deposit compositions (Lippman and Bunker, 1972; Gilluly, 1971; Dickinson and Hatherton, 1967; Sillitoe, 1972b). If a generalization may be made it is that the present distribution of gold deposits is a function of the Earth's behavior during the past 4.6 b. y. More specifically, the Earth's behavior appears to have determined not only the types of gold deposits but also when and where the deposits formed.

In order to understand the present distribution of gold deposits as a function of the Earth's behavior during the past 4.6 b. y., it is well to review the spatio-temporal relationships associated with gold deposits and the enclosing rock units. As indicated in earlier chapters, gold occurs mainly as native metal, always alloyed with various amounts of silver and other metals, mainly copper and iron. Besides native gold, the only important gold minerals are the tellurides, comprising a combination of gold, silver, copper, and (or) lead combined with tellurium. The most common tellurides are calaverite ($Au\ Te_2$) and sylvanite ($(Au, Ag)\ Te_2$). Gold also occurs in sulfide minerals.

When the global distribution patterns of gold deposits are studied over a long period of time, the first impression is that, although gold occurs in a limited number of forms, the deposits appear to be found in such a great variety of rock units as to almost preclude generalizations about the association of gold with particular rock types. Closer scrutiny reveals, however, that gold deposits occur more commonly with felsic or intermediate igneous rocks than with mafics and ultramafics. Also, the gold deposits occur more commonly in siliceous and aluminous sedimentary and metamorphic rocks than in carbonates. There are some exceptions, such as the new type exemplified by the Nevada gold deposits in silty dolomitic limestone at Carlin and Cortez.

Any attempt to place gold deposits into a fixed number of types is more-or-less arbitrary. But, historically, the most productive general types of deposits have been gold-quartz lodes, related epithermal bonanza deposits, geologically younger placers, fossil placers, marine placers, disseminated gold in bedrock, and gold as a byproduct of other minerals. It is well to consider each of these types of deposits in relation to the Earth's behavior during the past 4.6 b. y.

Gold-Quartz Lodes.

The gold-quartz lodes comprise a wide variety of essentially hydrothermal veins of quartz and gold that either replace wallrock (e. g., Homestake) or fill open spaces along fracture zones (e. g., Mother Lode). The lodes appear to have deep-seated origins because (1) they persist to great depths and (2) most are Precambrian ores in Precambrian rocks. Seismotectonomagmatic-belt activity during the Proterozoic and Phanerozoic has altered many of the "original" Archean gold deposits. This alteration spans a complete spectrum from slight wallrock alteration (e. g., Homestake) to widespread and deep-seated alteration (e. g., Alaska Juneau).

The size of the gold-quartz lodes ranges from more than 20 million ounces of gold (e. g., Homestake and Kolar) to others that are almost barren. In between are many lodes that have produced between 1 and 20 million ounces of gold: North Star, Empress, Alaska Juneau, Brownhill-Oroya, Bendigo, Ballarat, Hollinger, McIntyre, Dome, Kirkland Lake, Lake Shore, Kerr-Addison, and Morro Velho.

Homestake is the world's most productive gold lode mine, having produced over 30 million ounces of gold since its discovery in 1876. In this connection, many placers (such as Witwatersrand) produce almost that much gold per year (Brobst and Pratt, 1973). The lodes have probably produced about 1/4 of the world's gold output since 1900. Except for Alaska Juneau - Treadwell, Mother Lode - Grass Valley, and Bendigo - Ballarat, the major gold lodes are primarily Precambrian metamorphosed volcanic and sedimentary rocks of the greenstone-granite association, such as would be expected to result from seismotectonomagmatic-belt activity during the Archean and Proterozoic.

Epithermal Bonanza Deposits.

Most bonanza deposits are hydrothermal veins of quartz, carbonate minerals, barite, and fluorite, containing gold or gold tellurides and silver generally in excess of the gold content. The bonanzas appear to have been emplaced primarily by the filling of open spaces. Almost all of them are in highly altered Tertiary volcanic rocks. They are shallow deposits that are slightly richer than are the quartz lodes, ranging from a fraction of an ounce to 20 ounces per ton (e. g., Goldfield).

Among the largest bonanzas are Cripple Creek, Comstock, Goldfield, Telluride, Tonopah, Waihi, Thames, and El Oro. Others include those at Virginia City, Silverton, Ouray, Pachuca, Guanajuato, Transylvania (Romania), Nicaragua, and Japan. Very little gold is produced from bonanzas today. Most of these easy-to-find deposits have been worked out. They are probably the most easily discoverable bedrock gold deposits.

Bonanzas, according to the seismotectonomagmatic-belt concept, are vestiges of "original" Archean gold that has been repeatedly "reworked" during the Proterozoic and Phanerozoic, including particularly orogenic activity associated with the Tertiary seismotectonomagmatic-belts.

Young Placers.

The Earth's young placer gold deposits comprise mainly unconsolidated or semi-consolidated sand and gravel containing very small amounts of native gold and other heavy metals. Most young placers are stream deposits and occur along present stream valleys and on benches and terraces of former streams. A few are beach deposits (e. g., Nome) or residual deposits in deeply-weathered rock (e. g., eastern USA).

Even the richest young placers contain only a few grams of gold per ton of gravel. Productionwise, even the largest of these have yielded only a few million ounces of gold. Typical young placers include those along the American, Feather, and Yuba rivers in the Sierra Nevada of California; at Fairbanks, Alaska; on and near the beach at Nome, Alaska; at Klondike in the Yukon; in the upper reaches of the Siberian rivers, Lena, Amur, and Yenisei; along the east flank of the Andes of Bolivia, Colombia, Ecuador, and Peru; at Ballarat, Australia; in the conglomerates of northwestern Wyoming; in the Boise Basin, Idaho; along the Alder Gulch at Virginia City, Montana; in Egypt; in Spain; in Brazil; and in other areas that are associated with seismotectonomagmatic-belt activity that occurred during the late Cretaceous and much of the Tertiary.

Because these young placers were easy to find and easy to mine, most of them had been fairly well worked out by the 19th century.

Their total contribution to world gold production could have been as much as 1 billion ounces, or as much as 1/3 of the world's total production to date. Today, a significant amount of young-placer gold is produced only in Bolivia and the USSR, amounting to roughly 8% of world production.

Fossil Placers.

Fossil placers, mostly dating back to the Precambrian, were formed during Proterozoic seismotectonomagmatic-belt activity. Since Precambrian times, they have been little affected by subsequent seismotectonomagmatic-belt activity. Consequently, they now appear as part of the bedrock, mostly lithified to conglomerates. These comprise (1) small well-rounded quartz pebbles embedded within a matrix of pyrite and micaceous minerals and (2) a large suite of heavy minerals, including gold, uraninite, and the platinum-group metals.

The conglomerates associated with the fossil placers lay on the surface of the Earth prior to the genetic seismotectonomagmatic-belt activity that produced the fossil deposits. This activity served to bury and tilt the erstwhile flat-lying comglomerate beds. These tilted beds are only about a meter thick, but their planar extent runs into many km. The gold content ranges from about 0.2 to 0.8 oz/tn. Typical examples include Tarkwa (0.2 oz/tn), Witwatersrand (0.35 oz/tn), and Jacobina (0.5 oz/tn). The Witwatersrand district is the principal source of the world's gold. No fossil placer district approaching Witwatersrand has yet been discovered; but, based on analyses using the seismotectonomagmatic-belt concept, some very inviting prospects remain to be discovered, particularly in the Canadian, Brazil-Guyanan, and Africa-Arabian ancient shields. Some, but not all, of these prospects may be uneconomical to mine with present techniques.

Marine Placers.

The largest known auriferous area on the sea floor is in the northern Bering Sea. The gold concentration is less than 0.1 ppm and it occurs over a very broad area, including Norton Sound along the southern edge of Seward Peninsula. Similar concentrations have been found on the continental shelves off Oregon and northern California. Marine stanniferous placers are mined on a fairly large scale in Malaysia, Thailand, and Indonesia; these contain small (almost negligible) amounts of gold. The same may be said for the titaniferous minerals in Ceylon and Australia.

According to the seismotectonomagmatic-belt concept, marine placers resulted primarily from erstwhile land-based deposits that were emplaced with clastic sediments on the ocean floor after being transported by streams.

Disseminated Gold in Bedrock.

Most deposits of gold disseminated in bedrock are of relatively recent discovery, dating from about 40 years ago and younger (e. g., Cortez was discovered in 1966). The deposits comprise very fine grained gold (less than 10 microns) disseminated in silty and carbonaceous dolomitic limestone. The host rock is usually early Paleozoic (e. g., Carlin and Cortez are in a Silurian host).

The disseminated gold is accompanied by silica, some pyrite and other sulfides, and barite. The ore appears to have been formed by hydrothermal replacement of the host rock. The yield is about 0.3 oz/tn plus silver running about 10% that of the gold. Characteristic trace elements include arsenic, antimony, mercury, and sometimes tungsten.

Most of these deposits of disseminated gold were produced during seismotectonomagmatic-belt activity of the late Cretaceous and the Tertiary. Because of their recent discovery, very little is known about these deposits, but Carlin and Cortez contain more than a million oz of gold (Brobst and Pratt, 1973). Getchell has yielded nearly 1/2 million oz. At the present time, Carlin and Cortez account for roughly 1/4 of United States production. But, until more disseminated-gold deposits are discovered in other parts of the world, this type of deposit will contribute very little to world production.

The geologic environment of these deposits is not completely understood. However, there is a marked geochemical similarity between deposits and some modern hot-spring deposits in certain parts of the world. The hot-spring deposits of New Zealand contain as much as 85 ppm gold and 500 ppm silver (See, e. g., Weissberg, 1969). An initial analysis, using the seismotectonomagmatic-belt concept, indicates that all continents contain some highly inviting prospects of gold disseminated in late Mesozoic and Cenozoic rocks. Most of these are untested, but they include such well-known vestiges of seismotectonomagmatic-belt activity as the Antler orogenic belt of Nevara (Tatsch, 1975e).

Deposits that resemble Carlin, Cortez, and Getchell include Gold Acres (See, e. g., Brobst and Pratt, 1973; Dickson et al., 1975; Joralemon, 1975; Roberts et al., 1971; Wells and Mueller, 1973).

Byproduct Gold.

Despite the very low gold content in base metal ores (i. e., less than 1 ppm, or about 0.03 oz/tn), gold production as a byproduct is significant because of the large tonnages of base metals mined. Of the byproduct production, about 80% is from copper ore, with the remainder from complex ores of copper, lead, and zinc. Roughly 10% of the world's total gold production is from the byproduct source. Percentages vary from country to country. The highest percentage is probably in the United States, where byproduct gold accounts for 40% of the total production. Other countries where the precentages exceed 15% are Australia, Canada, Japan, Nicaragua, Peru, and the Philippines.

According to the seismotectonomagmatic-belt concept, the origin, evolution, and present characteristics of byproduct gold sources are largely the same as those for copper deposits (Tatsch, 1975a).

Chapter 12

THE TECTONOSPHERIC EARTH MODEL AS A SUPPLEMENTARY TOOL IN THE EXPLORATION FOR GOLD DEPOSITS

Four fundamental questions may be asked prior to exploring for gold: (1) what are the basic types of gold deposits; (2) which of these types offer the most gold at the least cost; (3) where are the best places to look for the types identified in (2); and (4) how can one most economically find the deposits identified in (3)?

The first of these questions has been discussed in Chapters 4 through 10 and summarized in Chapter 11. The other three questions are discussed in this chapter.

The Basic Types of Gold Deposits.

The basic types of gold deposits have been described in Chapter 11. They may be listed here:

a. Gold-quartz lodes.

b. Epithermal bonanza deposits.

c. Young placer.

d. Fossil placers.

e. Marine placers.

f. Disseminated gold in bedrock.

The origin, evolution, and present characteristics of these types of gold deposits have been summarized in Chapter 11.

The Basic Economics of Finding Hidden Gold Deposits.

It is axiomatic that the easiest-to-find gold is usually the first to be found. Most placers and near-surface bonanzas are relatively easy to find and easy to mine. Consequently, the known placers and bonanzas have been fairly well worked out. As the price of gold increases, various re-assessments must be made in connection with future prospects, particularly regarding placers in some areas.

The largest potential sources of gold in the United States, for example, are in the Cretaceous and Tertiary placers, but at the present time they are not economic to mine. The auriferous conglomerates of Wyoming contain 35 to 200 ppb gold in a volume exceeding 200 km^3. These conglomerates could easily contain several hundred million ounces of gold (Brobst and Pratt, 1973). The Tertiary placers of the Sierra Nevada may contain 10% that much. But legal and technological problems dictate against the feasibility of mining these placers now or within the foreseeable future.

Disseminated gold in bedrock (i. e., Carlin and Cortez types) offers a good possible source for future exploitation. At the present time, however, the geological and geochemical nature of these disseminations poses a problem. Although much is known about the the origin, evolution, and present characteristics of the Carlin-type deposits, no new deposits of this type have been found during the past 10 years.

The lode and bonanza type deposits promise to provide some good future prospects. However, most of those discoverable with present techniques appear to have been found. The complex geology of the Precambrian structures in which the lodes are found defies unravelling by present techniques. The complex geochemistry of the halos associated with bonanzas likewise defies interpretation when using present techniques.

The Most Favorable Areas for Future Exploration.

When all the legal, technological, and economic problems are considered, it appears that the most favorable prospects for future exploration are associated with the two types of deposits that have provided most of the gold deposits described in Chapters 6 through 10. These are the gold-quartz lodes and the fossil placers. These two types are probably vestiges of "original" Archean gold that has been the least reworked by late Archean and subsequent seismotectonomagmatic-belt activity. Most of the highly-productive gold-quartz lodes and fossil placers date back to the late Archean (2.7 to 2.6 b. y. ago) and early Proterozoic (2.6 to 1.6 b. y. ago).

Rocks of the late Archean and early Proterozoic are hosts for all the major Precambrian gold deposits of the world. These include the well-known deposits of Homestake, Kalgoorlie, Kirkland Lake - Larder Lake, Kolar, Morro Velho, Porcupine, Tarkwa, Siberia, and Witwatersrand. These are identified with the major shield areas of the world. Simple predictions suggest that hidden gold-quartz lodes and fossil placers lie within other areas of the Africa-Arabian, Australian, Baltic, Brazil-Guayanan, Canadian, Indian, and Siberian shields of late Archean and early Proterozoic age.

Although it seems logical that these old shield areas contain undiscovered deposits, recent research has not been too productive. Extensive exploration has failed to yield any deposits that even remotely resemble the gold-uranium placers of Witwatersrand.

Does this mean that there was a critical, undecipherable set of geometrical, mechanical, thermal, and chemical factors that contributed to the origin, evolution, and present characteristics of the Rand deposit? Has this set of factors operated anywhere else besides Witwatersrand during the "reworking" of the "original" Archean gold during the period 2.7 to 1.6 b. y. ago? The next section explores how these and other questions may be answered when applying the seismotectonomagmatic-belt concept to the exploration for gold deposits.

The Seismotectonomagmatic-Belt Concept Applied to the Exploration for Gold Deposits.

Because most gold deposits that display easily recognizable surficial manifestations have already been found, the future success in gold exploration will depend increasingly upon a more sophisticated approach, preferably one that embodies a clear understanding of the genetic relationships that exist between mineral deposits and geodynamic processes and phenomena. One such approach has been used in this book: it is based on the observation that economic concentrations of minerals are not randomly distributed but appear to be related to tectonic processes and regimes within the Earth. The details of these genetic relations are not known, and speculation regarding them has long been the subject of discussions among geologists. The resulting diversity of opinion can be seen by reading any applicable journal or by attending any meeting devoted to economic geology.

In the first 11 chapters of this book, an attempt has been made to determine how the Earth's gold deposits have originated, have evolved, and have become emplaced into their present distribution patterns in accordance with a single, long-lived, deep-seated, global mechanism that has been operating within the Earth during the 4.6 b. y. that the Earth is believed to have been in existence. This mechanism, the Tectonospheric Earth Model, permits the prediction of certain variations that have occurred within the geometrical, mechanical, thermal, and chemical heterogeneities of the Earth's upper 1000 km during the past 4.6 b. y. The ability of the model to provide these predictions forms the basis for its use as a supplementary tool in the exploration for gold deposits.

REFERENCES

Abbott, D. M., Jr., and Tautwein, C. M., 1975. Recognition of seven periods of Precambrian folding in the Colorado Front Range. Geol. Soc. Amer. Abstr. Prog., 7, 585.

Adaima, Sh. A., 1975. Plate tectonics and the development of the Alpine system: Discussion. Geol. Soc. Amer. Bull., 86, 719-720.

Ahrens, L. H. (Editor), 1968. Origin and Distribution of the Elements. Pergamon, New York, 1178 pp.

Ahrens, T. J., 1973. Petrologic properties of the upper 670 km of the Earth's mantle; geophysical implications. Phys. Earth Planet. Int., 7, 167-186.

Ahrens, T. J., and Schubert, G., 1975. Rapid formation of eclogite in a slightly wet mantle. Earth Planet. Sci. Ltr., 27, 90-94.

Allsman, P. T., 1940. Reconnaissance of gold-mining districts in the Black Hills, South Dakota. U. S. Bu. Min. Bull., 427, 147 pp.

Ames, R. L., 1962. The origin of gold-quartz deposits, Yellowknife. Econ. Geol., 57, 1137-1140.

Amstutz, G. C., and Bernard, A. J. (Editors), 1973. Ores in Sediments. Springer-Verlag, Berlin, 350 pp.

Anasenko, A. V., 1970. Gold deposits of the Adycha-Indigirka region. Geol. Ore Dep., 12(4), 105-108.

Anderson, A. L., 1942. The Boise Basin, Idaho. In: Newhouse (Ed.), (q. v.), pp. 132-134.

Anderson, T. A., 1975. Carboniferous subduction complex in the Harz Mountanns, Germany. Geol. Soc. Amer. Bull., 86, 77-82.

Anderson, T. H., and Silver, L. T., 1974. Late Cretaceous plutonism in Sonora, Mexico, and its relationship to Circum-Pacific magmatism. Geol. Soc. Amer. Abstr. Prog., 6, 484.

Anderson, R. N., Clague, D. A., Klitgord, K. D., Marshall, M., and Nishimori, R. K., 1975. Magnetic and petrologic variations along the Galapagos spreading center and their relation to the Galapagos melting anomaly. Geol. Soc. Amer. Bull., 86, 683-694.

Anguita, F., and Hernan, F., 1975. A propagating fracture model versus a hot spot origin for the Canary Islands. Earth Planet. Sci. Ltr., 27, 11-19.

Anhaeusser, C. R., Mason, R., Viljoen, M. J., and Viljoen, R. P., 1969. A reappraisal of some aspects of Precambrian shield geology. Bull. Geol. Soc. Amer., 80, 2175-2200.

Annels, A. E., 1974. Some aspects of the stratiform ore deposits of the Zambian Copperbelt and their genetic significance. In: Bartholome (Editor) (q. v.), pp. 235-254.

Anon., 1967. Witwatersrand gold: the challenge of inflation. World Min., 3(2), 12-17.

Anon., 1975. Val d'Or intersection encourages Belmoral. Nor. Min., 61(15), 23.

Anoshin, G. N., and Kepezhinskas, V. V., 1973. Petrochemical features related to gold distribution for the Cenozoic volcanic rocks of the Kuril-Kamchatka province. Geochem. Internat., 9, 618-628.

Antweiler, J. C., Doe, B. R., and Delevaux, M. H., 1972. Lead isotope and other evidence on the bedrock source of placer gold at Hahns Peak, Colorado. Econ. Geol., 67, 302-314.

Anufriyev, Yu. N., Stupakov, G. P., and Moshalyuk, A. A., 1974. Characteristics of quartz of Kochkar gold deposit. Internat. Geol. Rev., 16, 405-407.

Armstrong, R. L., 1975. Episodic volcanism in the central Oregon Cascade Range: Confirmation and correlation with the Snake River Plain. Geology, 3, 356-357.

Arth, J. G., and Hanson, G. N., 1975. Geochemisty and origin of the early Precambrian crust of northeastern Minnesota. Geochim. Cosmochim. Acta, 39, 325-362.

Artyushkov, E. V., 1974. Can the Earth's crust be in a state of isostasy: J. Geophys. Res., 79, 741-752.

Atabekyants, K. P., 1973. Zonation in primary halos of Yuzhnoye Island auriferous ores, as influenced by formation characteristics of the deposit. Internat. Geol. Rev., 15, 1394-1403.

Athavale, R. N., 1973. Inferences from recent Indian paleomagnetic results about the northern margin of the Indian plate and the tectonic evolution of the Himalayas. In: Tarling and Runcorn (Editors), (q. v.), vol. 1, pp. 117-130.

Atwater, T., 1973. Studies of sea-floor spreading and plate movements in the Pacific basin. In: Tarling and Runcorn (Editors) (q. v.), vol. 1, pp. 213-217.

Aubouin, J., 1965. Geosynclines. Elsevier, Amsterdam, 355 pp.

Audley-Charles, M. G., 1975. The Sumba fracture: A major discontinuity between eastern and western Indonesia. Tectonophys., 26, 213-228.

Azhgirei, G. D., 1968. Folding and mountain building (orogenesis). In: Mistik (Editor) (q. v.), pp. 287-298.

Badham, J. P. N., and Halls, C., 1975. Microplate tectonics, oblique collisions, and evolution of the Hercynian orogenic systems. Geology, 3, 373-376.

Baer, A. J.,(Editor), 1970. Basins and Geosynclines of the Canadian Shield. Geological Survey of Canada, Ottawa, 265 pp.

Baer, A. J., 1975. The Grenville belt. Mtg. Roy. Soc. Lond., Mar. 13-14.

Bain, G. W., 1968. Syngenesis and epigenesis of ores in layered rocks. In: Stemprok (Editor) (q. v.), pp. 119-136.

Baird, A. K., Morton, D. M., Woodford, A. O., and Baird, K. W., 1974. Transverse Range Province: A unique structural-petrochemical belt across the San Andreas fault system. Geol. Soc. Amer. Bull., 85, 163-174.

Baker, B. H., Mohr, P. A., and Williams, L. A. J., 1972. Geology of the Eastern Rift System of Africa. Geological Society of America, Boulder, 67 pp.

Banerjee, P. K., and Gosh, S., 1972. Correlation of Precambrian ore provinces of east Africa, India, and West Australia. Econ. Geol., 67, 55-62.

Barazangi, M., and Dorman, J., 1969. World seismicity map of ESSA Coast and Geodetic Survey epicenter data for 1961-1967. Bull. Seis. Soc. Amer., 59, 369-380.

Barazangi, M., Pennington, W., and Isacks, B., 1975. Global study of seismic wave attenuation in the upper mantle behind island arcs using pP waves. J. Geophys. Res., 80, 1079-1092.

Barnes, H. L., (Editor), 1967. Geochemistry of Hydrothermal Ore Deposits. Holt, Rinehart, and Winston, New York, 584 pp.

Bartholome, P., 1974a. On the diagenetic formation of ores in sedimentary beds, with special reference to Kamoto, Shaba, Zaire. In: Bartholome (Editor) (q. v.), pp. 203-213.

Bartholome, P. (Editor), 1974b. Gisements Stratiformes et Provinces Cupriferes. Soc. Geol. Belg., Liege, 427 pp.

Bateman, A. M., 1950. Economic Mineral Deposits, 2nd ed. Wiley, New York, 916 pp.

Bateman, A. M., 1951. The Formation of Mi eral Deposits. Wiley, New York, 371 pp.

Bateman, J. D., 1958. Uranium-bearing auriferous reefs at Jacobina, Brazil. Econ. Geol., 53, 417-425.

Bayley, R. W., 1970. Structure and mineralization of Precambrian rocks in the Galena-Roubaix district, Black Hills, South Dakota. U. S. Geol. Surv. Bull. 1312-E, 15 pp.

Beloussov, V. V., 1962. Basic Problems in Geotectonics. McGraw-Hill, New York, 816 pp.

Beloussov, V. V., 1968. Some general aspects of development of the tectonosphere. In: Benes (Editor) (q. v.), pp. 9-17.

Beloussov, V. V., 1969a. Continental rifts. In: Hart (Editor)

(q. v.), pp. 539-544.

Beloussov, V. V., 1969b. Interrelations between the Earth's crust and upper mantle. In: Hart (Editor) (q. v.), pp. 698-712.

Beloussov, V. V., 1970. Against the hypothesis of ocean floor spreading. Tectonophys., 9, 489-511.

Beloussov, V. V., 1971. On possible forms of relationship between magmatism and tectonogenesis. J. Geol. Soc. 127, 57-68.

Beloussov, V. V., 1972. Important trends in the evolution of continents. Internat. Geol. Cong. Abstr., 24, 65.

Benes, K. (Editor), 1968. Upper Mantle (Geological Processes). Academia, Prague, 260 pp.

Benioff, H., 1959. Circum-Pacific tectonics. Publs. Domin. Obs. Ottawa, 20, 395-402.

Bennett, G. T., Clowes, R. M., and Ellis, R. M., 1975. A seismic refraction survey along the southern Rocky Mountain Trench, Canada. Bull. Seis. Soc. Amer., 65, 37-54.

Berg, E., and Sutton, G. H., 1975. Dynamic interaction of seismic and volcanic activity of the Nazca plate edges. Phys. Earth Planet. Int., 9, 175-182.

Best, M. G., and Brimhall, W. H., 1974. Late Cenozoic alkalic basaltic magmas in the western Colorado plateaus and the Basin and Range transition zone, U. S. A., and their bearing on mantle dynamics. Geol. Soc. Amer. Bull., 85, 1677-1690.

Bichan, J. W., 1947. Structural principles controlling occurrence of ore in Kolar Gold Field, India. Econ. Geol., 42, 93-136.

Bickford, M. E., and Van Schmus, W. R., 1973. Possible middle and late Precambrian igneous arcs in the mid-continent region of North America. Geol. Soc. Amer. Abst. Prog., 5, 300.

Blissenbach, E., and Fellerer, R., 1973. Continental drift and the origin of certain mineral deposits. Geol. Rund., 62, 812-839.

Boles, J. R., and Coombs, D. S., 1975. Mineral reactions in zeolitic Triassic tuff, Hokonui Hills, New Zealand. Geol. Soc. Amer.

Bull., 86, 163-173.

Bonatti, E., 1975. Metallogenesis at oceanic spreading centers. Ann. Rev. Earth Planet Sci., 3, 401-431.

Borrello, A. J., 1972. The Precordillera as a type of geosyncline in Argentina. Internat. Geol. Cong. Abstr., 24, 67.

Bostrom, K., Farquharson, B., and Eyl, W., 1972. Submatine hot springs as a source of active ridge sediments. Chem. Geol., 10, 189-198.

Bowes, D. R., 1975. Tectonics in the Baltic Shield in the period 2000-1500 million years. Mtg. Roy. Soc. Lond., Mar. 13-14.

Bowin, C., 1974. Migration of a pattern of plate motion. Earth Planet. Sci. Ltr., 21, 400-404.

Boyle, R. W., 1959. The geochemistry, origin, and role of carbon dioxide, water,,sulfur, and boron in the Yellowknife gold deposits, Northwest Territories, Canada. Econ. Geol., 54, 1506-1524.

Boyle, R. W., 1961. The geology, geochemistry, and origin of the gold deposits of the Yellowknife district. Can. Geol. Surv. Mem. 310, 193 pp.

Boyle, R. W., Wanless, R. K., and Lowden, J. A., 1963. The origin of the gold-quartz deposits, Yellowknife, N. W. T. Econ. Geol., 58, 804-807.

Brobst, D. A., and Pratt, W. P. (Editors), 1973. United States Mineral Resources. Govt. Printing Off., Washington, 722 pp.

Brock, B. B., 1972. A Global Approach to Geology: The Background of a Mineral Exploration Strategy Based on Significant Form in the Patterning of the Earth's Crust. Balkema, Cape Town, 365 pp.

Brock, A., 1973. Precambrian polar wander and continental drift of Africa and North America: Some comments. In: Tarling and Runcorn (Editors) (q. v.), vol. 1, pp. 33-36.

Brown, B. W., 1974. Sediment accumulation, sea-floor spreading, and eustacy: Discussion. Geol. Soc. Amer. Bull., 85, 831.

Brown, M. R., 1975. Amoco gold discovery looks major. Nor. Min., 61(14), 1, 8, 20.

Browne, P. R. L., 1969. Sulfide mineralization in a Broadlands geothermal drillhole, Taupo volcanic zone, New Zealand. Econ. Geol., 64, 156-159.

Bruce, E. L., 1942. The mineral deposits of Thunder Bay district, Ontario. In: Newhouse (Editor) (q. v.), pp. 101-106.

Bullen, K. E., 1974a. Introductory remarks on standard Earth model. Phys. Earth Plnt. Int., 9, 1-3.

Bullen, K. E., 1974b. Standard Earth model requirements in respect of density and rigidity in the inner core. Phys. Earth Planet. Int., 9, 41-44.

Butler, B. S., 1942. Some interrelations of structure, mineralogy, and association with intrusive bodies in ore deposits. In: Newhouse (Editor) (q. v.), pp. 3-5.

Cabre, R., 1975. Geodynamics in the eastern Pacific and the western Americas. Phys. Earth Planet. Int., 9, 169-173.

Carey, W. S., 1975. The expanding Earth: An essay review. Earth Sci. Rev., 11, 105-143.

Cawthorn, R. G., 1975. Degrees of melting in mantle diapirs and the origin of ultrabasic liquids. Earth Planet. Sci. Ltr., 27, 113-120.

Cawthorn, R. G., and Strong, D. F., 1974. The petrogenesis of komatiites and related rocks as evidence for a layered upper mantle. Earth Planet. Sci. Ltr., 23, 369-375.

Cepeda, J. C., 1973. Evidence for multiple Precambrian deformation in the Sangre de Cristo Mountains, New Mexico. Geol. Soc. Amer. Abstr. Prog., 5, 470.

Chace, F. M., 1949. Origin of the Bendigo saddle reefs and formation of ribbon quartz. Econ. Geol., 44, 561-597.

Challinor, J., 1964. A Dictionary of Geology, 2nd ed. Oxford Univ. Press, New York, 289 00.

Chase, C. G., Herron, E. M., and Normack, W. R., 1975. Plate tectonics: Commotion in the ocean and continental sequences. Ann.

Rev. Earth Planet. Sci., 3, 271-291.

Churkin, M., 1975. Basement rocks of Barrow Arch, Alaska, and Circum-Arctic Paleozoic mobile belt. Amer. Assoc. Petrol. Geol. Bull., 59, 451-456.

Coleman, L. C., 1957. Mineralogy of the Giant Yellowknife gold mine, N. W. T. Econ. Geol., 52, 400-425.

Coleman, P. J., 1975. On island arcs. Earth Sci. Rev., 11, 47-80.

Colley, H., and Warden, A. J., 1974. Petrology of the New Hebrides. Geol. Soc. Amer. Bull., 85, 1635-1646.

Corliss, J. B., 1973. Volcanism and ore-formation related to the subduction of oceanic crust: the "geostill" concept. Geol. Soc. Amer. Abstr. Prog., 5, 26.

Cormier, V. F., 1975. Tectonics near the junction of the Aleutians and Kuril-Kamchatka arcs and a mechanism for middle Tertiary magmatism in the Kamchatka basin. Geol. Soc. Amer. Bull,, 86, 443-453.

Coulon, C., and Dupuy, C., 1975. Evolution spatiale des caracteres chimiques du volcanisme andesitique de la Sardaigne (Italie). Earth Planet. Sci. Ltr., 25, 170-176.

Coward, M. P., 1975. Archean deformation patterns in southern Africa. Mtg. Roy. Soc. Lond., Mar. 20-21.

Damon, P. E., 1972. Orogenic-epeirogenic periodicity. Internat. Geol. Cong. Abstr., 24, 72.

Dandurand, J. L., Fortune, J. P., Perami, R., Schott, J., and Tollon, F., 1972. On the importance of mechanical action and thermal gradient on the formation of metal-bearing deposits. Min. Dep., 7, 339-350.

Darracott, B. W., 1974. On the crustal structure and evolution of southeastern Africa and the adjacent Indian Ocean. Earth Planet. Sci. Ltr., 24, 282-290.

Dasch, E. J., 1975. Metallogenesis in the southeastern Pacific:

A progress report on the IDOE Nazca plate project. Phys. Earth Planet. Int., 9, 249-258.

Davidenko, N. M., 1970. Zoning of auriferous ore bodies in the Keperveem area of Chukot Peninsula. Geol. Ore Dep., 12(2), 103-106.

Davidson, C. F., 1960. The present state of the Witwatersrand controversy. Min. Mag., 102, 84-95, 149-159, 222-229.

Dean, R. L., Strong, D. F., 1975. The volcanic stratigraphy, geochemistry, and metallogeny of the central Newfoundland Appalachians. Geol. Soc. Amer. Abstr. Prog., 7, 745-746.

DeKun, N., 1965. The Mineral Resources of Africa. Elsevier, Amsterdam, 740 pp.

Derry, D. R., 1942a. The Matachewan consolidated mine. In: Newhouse (Editor) (q. v.), pp. 183-187.

Derry, D. R., 1942b. The Canadian Malarctic gold mine. In: Newhouse (Editor) (q. v.), pp. 247-248.

Deryabin, N. I., 1970. Structure of fault zones in the Maslovskiy deposits in the northern Urals. Geol. Ore Dep., 12(4), 41-55.

Dickinson, W. R., and Hatherton, T., 1967. Andesitic volcanism and seismicity around the Pacific. Science, 157, 801-803.

Dickson, F. W., Radtke, A. S., Weissberg, B. G., and Heropoulos, C., 1975. Solid solutions of antimony, arsenic, and gold in stibnite (Sb_2S_3), orpiment (As_2S_3), and realgar (As_2S_2). Econ. Geol., 70, 591-594.

Dietz, R. S., 1952. Geomorphic evolution of continental terrace (continental shelf and slope). Bull. Amer. Assoc. Petrol. Geol., 36, 1802-1819.

Dietz, R. S., 1963a. Wave base, marine profile of equilibrium, and wave-built terraces: a critical appraisal. Geol. Soc. Amer. Bull., 74, 971-990.

Dietz, R. S., 1963b. Collapsing continental rises: an actualistic concept of geosynclines and mountain building. J. Geol., 71, 314-333.

Dietz, R. S., and Holden, J. C., 1966. Miogeosynclines in space and time. J. Geol., 74, 566-583.

Dietz, R. S., and Holden, J. C., 1970. East Indian basin (Wharton Basin) as pre-Mesozoic ocean crust. Geol. Soc. Amer., Abstr. Prog., 2, 537.

Dietz, R. S. and Holden, J. C., 1973. Continental drift: New orthodoxy or persuasive joker? In: Tarling and Runcorn (Editors) (q. v.), vol. 2, pp. 1105-1121.

Dietz, R. S., and Menard, W. H., 1951. Origin of abrupt change in slope at continental shelf margin. Amer. Assoc. Petrol. Geol. Bull., 35, 1994-2016.

Dietz, R. S., Holden, J. C., and Sproll, W. P., 1969. Geotectonic evolution and subsidence of Bahama platform. Geol. Soc. Amer. Abst. Prog., 7, 48.

Dimitriyev, L. V., 1972. Possible petrological consequences of the rise of mantle beneath mid-ocean ridges. Geochem. Internat., 1972, 168-172.

Dimroth, E., Baragar, W. R. A., Bergeron, R., and Jackson, G. D., 1970. The filling of the Circum-Ungava geosyncline. Geol. Surv. Can., Paper 70-40, 45-142.

Dixon, C. J., Pereira, J., 1973. Plate tectonics and mineralization in the Middle East. Econ. Geol., 68, 137.

Dixon, C. J., and Pereira, J., 1974. Plate tectonics and mineralization in the Tethyan region. Min. Dep., 9, 185-198.

Doe, B. R., and Smith, D. K. (Editors), 1972. Studies in Mineralogy and Precambrian Geology. Geol. Soc. Amer. Mem. 15, 348 pp.

Doe, B. R., and Stacey, J. S., 1974. The applicztion of lead isotopes to the problems of ore genesis and ore prospect evaluation: A review. Econ. Geol., 69, 757-776.

Donath, F. A., Stehli, F. G., and Wetherill, G. W., 1975. Annual Review of Earth and Planetary Sciences. Annual Reviews, Palo Alto, 466 pp.

Dott, R. H., Jr., and Shaver, R. H. (Editors), 1974. Modern and Ancient Geosynclinal Sedimentation. Amer. Assoc. Petrol. Geol., Tulsa, 380 pp.

Douglas, R. J. W. (Editor), 1970. Geology and Economic Minerals of Canada. Geol. Surv. Can., Ottawa, 838 pp.

Drake, C. L., 1970. A long-range program of solid Earth studies. Trans. Amer. Geophys. Union, 51, 152-159.

Drake, C. L. (Editor), 1973. U. S. Program for the Geodynamics Project: Scope and Objectives. Nat. Acad. Sci., Washington, 235 pp.

Dubois, J., Launay, J., and Recy, J., 1975. Some new evidence on lithospheric bulges close to island arcs. Tectonophys,, 26, 189-196.

Dunnet, D., 1975. Precambrian deformation patterns in Australia. Mtg. Roy. Soc. Lond., Mar. 20-21.

D'Yachenko, A. P., Legendza, V. Ya., Shelekhov, A. N., and Radionova, Ye. S., 1970. Occurrence of auriferous ore in Precambrian rocks of the Staryy Oskol region of the Kursk Magnetic Anomaly. Geol. Ore Dep., 12(3), 92-96.

Dymond, J., 1975. K-Ar ages of Tahiti and Moorea, Society Islands, and implications for the hot-spot models Geology, 3, 236-240.

Dzotsenidze, G. S., 1969. Role of Volcanism in the Formation of Sedimentary Rocks and Ores. Nyedra, Moscow, 344 pp.

Eardley, A. J., 1963. Relation of uplifts to thrusts in Rocky Mountains. In: The Backbone of the Americas: Tectonic History from Pole to Pole. Amer. Assoc. Petrol. Geol. Mem. 2.

Ebbutt, F., 1942. The Britannia Mines, B. C., In: Newhouse (Editor) (q. v.), pp. 155-156.

El-Boushi, I. M., 1972. Geology of the Gebeit gold mine, Democratic Republic of Sudan. Econ. Geol., 67, 481-486.

Emmons, W. H., 1937. Gold Deposits of the World. McGraw-Hill, New York, 562 pp.

Ernst, W. G., 1975. Systematics of large-scale tectonics and age progressions in Alpine and Circum-Pacific blueschist belts. Tectonophys., 26, 229-246.

Escher, A., Jack, S., and Watterson, J., 1975. Brittle deformation, dyke intrusion, and the mechanical characteristics of Early Proterozoic crust in the North Atlantic region. Mtg. Roy. Soc. Lond., Mar. 13-14.

Evans, A. M., 1975. Mineralization in geosynclines: The Alpine enigma. Min. Dep., 10, 254-260.

Faure, H., 1972. Paleodynamique du craton africain. Internat. Geol. Cong. Abstr., 24, 75.

Fein, C. D., 1973. Chemical composition and submarine weathering of some oceanic tholeiites from the East Pacific Rise. Geol. Soc. Amer. Abstr. Prog., 5, 41.

Folinsbee, R. E., Kirkland, K., Nekolaichuk, A., and Smejkal, V., 1972. Chinkuashih: a gold-pyrite-enargite-barite hydrothermal deposit in Taiwan. In: Doe and Smith (Editors) (q. v.), pp. 323-335.

Froidevaux, C. and Schubert, G., 1975. Plate motion and structure of the continental asthenosphere: A realistic model of the upper mantle. J. Geophys. Res., 80, 2553-2564.

Gair, J. E., 1962. Geology of ore deposits of the Nova Lima and Rio Acima quadrangles, Minas Gerais, Brazil. U. S. Geol. Surv. Prof. Ppr. 341-A, 67 pp.

Gapontsev, G. P., 1971. Determination of opening depth of gold-bearing veins of the Baley-type according to geochemical data. Geol. Ore Dep., 13(1), 106-110.

Gapontsev, G. P., Polikarpochkin, V. V., Sarapulova, V. N., and Nesterova, A. A., 1969. The distribution of companion elements of gold within the planes of veins in the ore-bearing field of

Baley. Geol. Ore Dep., 11(5), 104-109.

Gast, P. W., 1972. The chemical composition of the Earth, the Moon, and chronditic meteorites. In: Robertson (Editor) (q. v.), pp. 19-40.

Gastil, G., 1960. The distribution of mineral dates in time and space. Amer. J. Sci., 258, 1.

Gastil, R. G., 1975. Plutonic zones in the Peninsular Ranges of southern California and northern Baha California. Geology, 3, 361-363.

Gastil, R. G., and Krummenacher, D., 1975. Reconstruction of the Mesozoic plutonic belt of southern California and western Mexico. Geol. Soc. Amer. Abstr. Prog., 7, 320.

Gavrikova, S. N., and Larionov, A. M., 1970. The geological and the physiochemical features of the listwanite formation of the Kenterlau massif. Geol. Ore Dep., 12(5), 41-51.

Gibson, R., 1942. The Libby Quadrangle, Montana. In: Newhouse (Editor) (q. v.), pp. 156-158.

Gill, J. 1973. Are island arc magmas derived from underthrust lithosphere? Trans. Amer. Geophys. Union, 54, 506.

Gill, R. C. O,, and Bridgwater, D., 1975. Geochemistry and significance of early Archean low-K and related tholeiite dykes cutting granitic gneisses in West Greenland. Geol. Soc. Amer. Abstr. Prog., 7, 763.

Gilluly, J., 1971. Plate tectonics and magmatic evolution. Geol. Soc. Amer. Bull., 82, 2383-2396.

Gilluly, J., 1972. Tectonics involved in the evolution of mountain ranges. In: Robertson (Editor) (q. v.), pp. 406-439.

Goodwin, A. M., 1968. Archean protocontinental growth and early crustal history of the Canadian shield. In: Benes (Editor) (q. v.), pp. 69-89.

Goodwin, A. M., 1973. Plate tectonics and evolution of Precambrian crust. In: Tarling and Runcorn (Editors) (q. v.), vol. 2, pp. 1047-1069.

Goodwin, A. M., 1975. Archean volcanism in Superior Province, Canadian Shield. Geol. Soc. Amer. Abstr. Prog., 7, 764.

Gottfried, D., Rowe, J. J., and Tilling, R. L., 1972. Distribution of gold in igneous rocks. U. S. Geol. Surv. Prof. Ppr. 727, 42 pp.

Graham, S. A., Dickinson, W. R., and Ingersoll, R. V., 1975. Himalayan-Bengal model for flysch dispersal in the Appalachian-Ouachita system. Geol. Soc. Amer. Bull., 86, 273-286.

Granath, J. W., 1975. Wind River Canyon: An example of a greenstone belt in the Archean of Wyoming, U. S. A. Precambr. Res., 2, 71-91.

Graton, L. C., 1930. Hydrothermal origin of the Rand gold deposits. Econ. Geol., 25, suppl. to no. 3.

Grebenchikov, A. M., and Sorokin, V. N., 1972. Endogene zoning and distribution of gold at the Taskorino deposit (Northern Lake Balkash region). Geol. Rudn. Mest., 14(6), 75-78.

Green, T., 1968. The World of Gold. Walker, New York, 242 pp.

Green, D. H., 1975. Genesis of Archean peridotite magmas and constraints of Archean geothermal gradients and tectonics. Geology, 3, 15-18.

Green, D. H., Nicholls, I. A., Viljoen, M., and Viljoen, R., 1975. Experimental demonstration of the existence of peridotitic liquids in earliest Archean magmatism. Geology, 3, 11-14.

Greenwood, W. R., Hadley, D. G., Anderson, R. E., Fleck, R. J., and Schmidt, D. L., 1975. Late Proterozoic cratonization in the southern part of the Arabian shield. Mtg. Roy. Soc. Lond., Mar. 13-14.

Gregorchuk, G. Yu., Ivasiv, S. M., and Lyakhov, Yu. V., 1971. Cyclic features of hydrothermal mineralization, using east Zabaikalye gold deposits as an example. Geol. Ore Dep., 13(5), 64-75.

Griffin, V. S., Jr., 1970. Relevancy of the Dewey-Bird hypothesis of cordilleran-type mountain belts and the Wegmann stockwork concept. J. Geophys. Res., 75, 7504-7507.

Griffin, V. S., Jr., 1974. Plate tectonics and the Ouachita system of Texas, Oklahoma, and Arkansas: Discussion. Geol. Soc. Amer. Bull., 85, 145-146.

Griggs, D. T., 1972. The sinking lithosphere and the focal mechanism of deep earthquakes. In: Robertson (Editor) (q. v.), pp. 361-384.

Gross, W. H., 1968. Evidence of a modified placer origin for auriferous conglomerates, Canavieiras mine, Jacobina, Brazil. Econ. Geol., 63, 271-276.

Grow, J. A., and Bowin, C. O., 1975. Evidence for high-density crust and mantle beneath the Chile trench due to the descending lithosphere. J. Geophys. Res., 80, 1449-1458.

Guild, P. W., 1973. Massive sulfide deposits as indicators of former plate boundaries. Econ. Geol., 68, 137-138.

Gunning, H. C., 1942. Gold deposits of Cadillac township, Quebec. In: Newhouse (Editor) (q. v.), pp. 163-165.

Gupka, H. K. (Editor), 1974. Seminar on the Geodynamics of the Himalayan Region. Natl. Geophys. Res. Inst., India, 221 pp.

Hackett, J. P., and Bischoff, J. L., 1973. New data on the stratigraphy, extent, and geologic history of the Red Sea geothermal deposits. Econ. Geol., 68, 553-564.

Hallberg, J. A., 1971. Geochemistry of Archean belts in the eastern Goldfields region of Western Australia. J. Petrol., 13, 45-54.

Hanson, G., 1942. The Barkerville Gold belt, British Columbia. In: Newhouse (Editor) (q. v.), pp. 167-177.

Harrison, J. E., 1972. Precambrian Belt basin of northwestern United States: its geometry, sedimentation, and copper ocuurrences. Geol. Soc. Amer. Bull., 83, 1215-1240.

Hart, P. J. (Editor), 1969. The Earth's Crust and Upper Mantle: Structure, Dynamic Processes, and Their Relation to Deep-Seated

Geological Processes. Amer. Geophys. Union, Washington, 735 pp.

Hattori, K., 1975. Geochemistry of ore deposition at the Yatani lead-zinc and gold-silver deposit, Japan. Econ. Geol., 70, 677-693.

Hawkesworth, C. J., Waters, D. J., and Bickle, M. J., 1975a. Plate tectonics in the eastern Alps. Earth Planet. Sci. Ltr., 24, 405-413.

Hawkesworth, C. J., Moorbath, S., O'Nions, R. K., and Wilson, J. F., 1975b. Age relationships between greenstone belts and "granites" in the Rhodesian Archean craton. Earth Planet. Sci. Ltr., 25, 251-262.

Hawkins, J. W., and Natland, J. H., 1975. Nephelinites and basanites of the Somoan linear volcanic chain: Their possible tectonic significance. Earth Planet. Sci. Ltr., 24, 427-439.

Hawley, J. E., 1942. Some gold mines of the Rouyn-Harricanaw belt, northwestern Quebec. In: Newhouse (Editor) (q. v.), pp. 95-101.

Hardie, B. S., 1966. Carlin gold mine, Lynn district, Nevada. Bu. Min. Rpt. 13A, 73-83.

Heirtzler, J. R. (Editor), 1972. Understanding the Mid-Atlantic Ridge. National Academy of Science, Washington, 131 pp.

Hester, B. W., 1972. Geology and evolution of placer deposits in the Klondike area, Yukon Territory. Trans. Inst. Min. Metal., 80, 57-59.

Hills, E. S., 1972. Time and space relationships of basins of epeirogenic origin in Australia. Internat. Geol. Cong. Abstr., 24, 80.

Hills, F. A., Houston, R. S., and Subbarayudu, G. V., 1975. Possible Proterozoic plate boundary in southern Wyoming. Geol. Soc. Amer. Abstr. Prog., 7, 614.

Hoffman, P. F., 1975. Proterozoic post-orogenic volcanism, Great Bear Lake, NWT. Geol. Soc. Amer. Abstr. Prog., 7, 785-786.

Hollister, V. F., Potter, R. R., and Barker, A. L., 1974. Porphyry-type deposits of the Appalachian orogen. Econ. Geol., 69, 618-630.

Holmes, A., 1965. Principles of Physical Geology. Ronald, New York, 1288 pp.

Houston, M. H., and DeBremaecker, J. Cl., 1975. Numerical models of convection in the upper mantle. J. Geophys. Res., 80, 742-751.

Hubbard, F. H., 1975. Precambrian crustal development in western Nigeria: Indications from the Iwo region. Geol. Soc. Amer. Bull., 86, 548-554.

Hurst, M. E., 1942. Gold deposits at Porcupine, Ontario. In: Newhouse (Editor), (q. v.), pp. 196-199.

Hutchinson, R. W., and Hodder, R. W., 1972. Possible tectonic and metallogenic relationships between porphyry copper and massive sulphide deposits. Can. Min. Metall. Bull., 65, 34-40.

Hyndman, D. W., Talbot, J. L., and Chase, R. B., 1975. Boulder batholith: A result of emplacement of a block detached from the Idaho batholith infrastructure? Geology, 3, 401-404.

Irving, E., and McGlynn, J. C., 1975. Precambrian paleomagnetism of the northern continents and its tectonic implications. Mtg. Roy. Soc. Lond., Mar. 13-14.

Irving, E., Emslie, R. F., and Ueno, H., 1974. Upper Proterozoic paleomagnetic poles from Laurentia and the history of the Grenville structural province. J. Geophys. Res., 79, 5491-5502.

Isacson, P. E., 1975. Evidence for a western extracontinental land source during the Devonian period in the central Andes. Geol. Soc. Amer. Bull., 86, 39-46.

Isihara, S., Igarashi, T., and Nishiwaki, C., 1975. A re-examination of the regional distribution of the late Cenozoic ore deposits in the East Japan Arc. Geol. Soc. Amer. Bull., 86, 292-294.

Jackson, E. D., and Shaw, H. R., 1975. Stress fields in central portions of the Pacific plate: Delineated in time by linear volcanic chains. J. Geophys. Res., 80, 1861-1874.

Jackson, E. D., Silver, E. A., and Dalrymple, G. B., 1972. Hawaiian Emperor chain and its relation to Cenozoic circimpacific tectonics. Geol. Soc. Amer. Bull,, 83, 601-618.

Jenks, W. F., 1975. Origins of some massive pyritic ore deposits of Western Europe. Econ Geol., 70, 488-498.

Jones, R. S., 1968. Gold in meteorites and in the Earth's crust. U. S. Geol. Surv. Circ. 603, 4pp.

Jones, R. S., 1969. Gold in igneous, sedimentary, and metamorphic rocks. U. S. Geol. Surv. Circ. 610, 28 pp.

Jones, R. S., 1970. Gold content of water, plants, and animals. U. S. Geol. Surv. Circ. 625, 15 pp.

Jones, R. S., and Fleischer, M., 1969. Gold in minerals and the composition of native gold. U. S. Geol. Surv. Circ. 612, 17 pp.

Joralemon, I. B., 1942. Bralorne, B. C., orebodies. In: Newhouse (Editor) (q. v.), p. 255.

Joralemon, P., 1975. K-Ar relations of granodiorite emplacement and tungsten and gold mineralization near the Getchell mine, Hulboldt County, Nevada. Econ. Geol., 70, 405-406.

Jordan, T. H., and Frazer, L. N., 1975. Crustal and upper mantle structure from Sp phases. J. Geophys. Res., 80, 1504-1518.

Juan, V. C., 1975. Tectonic evolution of Taiwan. Tectonophys., 26, 197-212.

Jurdy, D. M., and Van der Voo, R., 1975. True polar wander since the Early Cretaceous. Science, 187, 1193-1196.

Kanasewich, E. R., and Gutowski, P. R., 1975. Detailed seismic analysis of a lateral mantle inhomogeneity. Earth Planet. Sci. Ltr., 25, 379-384.

Kantsel, A. V., and Orlovskiy, N. B., 1971. Mathematical

modeling of ore deposits in relation to prognosis and investigations of formation conditions. Geol. Ore Dep., 13(6), 68-80.

Kanungo, S. C., 1972. Some aspects of gold mineralization in Koraput district, Arissa. Ind. Min., 26(2), 61-64.

Karig, D. E., and Sharman, G. F., III, 1975. Subduction and accretion in trenches. Geol. Soc. Amer. Bull., 86, 377-389.

Katili, J. A., 1975. Volcanism and plate tectonics in the Indonesian island arcs. Tectonophys., 26, 165-188.

Kato, T., 1942. Some ore deposits in Japan. In: Newhouse (Editor) (q. v.), pp 235-236.

Kaula, W. M., 1975. Absolute plate motions by boundary velocity minimizations. J. Geophys. Res., 80, 244-248.

Kavanagh, P. M., 1968. Have 6000 years of gold mining exhausted the world's gold reserves? Can. Min. Metall. Bull., 61, 553-558.

Kay, M., 1951. North American geosynclines. Geol. Soc. Amer. Mem. 48, 143 pp.

Keays, R. R., and Kirkland, M. C., 1972. Hydrothermal mobilization of gold from copper-nickel sulfides and ore genesis at the Thomson River Copper Mine, Victoria, Australia. Econ. Geol., 67, 1263-1275.

Kesler, S. E., 1973. Copper, molybdenum, and gold abundances in porphyry copper deposits. Econ. Geol., 68, 106-112.

Khain, V. E., 1972. Main trends in the evolution of the Earth's crust (lithosphere). Internat. Geol. Cong. Abstr., 24, 84.

Kieft, C., and Oen, J. S., 1973. Ore minerals of the telluride-bearing gold-silver ores of Salida, Indonesia, with special reference to the distribution of selenium. Min. Dep., 8, 312-320.

King, J. R., Bruce, W. R., Slater, M. N., and Field, C. W., 1975. The geology and mineral deposits of the Cuddy Mountains, West Central Idaho. Geol. Soc. Amer. Abstr. Prog., 7, 617.

Knopf, A., 1942. Ore deposition in the pyrometasomatic deposits. In: Newhouse (Editor) (q. v.), pp. 63-72.

Knopoff, L., and Fouds, A. A., 1975. Upper-mantle structure under the Arabian Peninsula. Tectonophys., 26, 121-134.

Kolkovsky, B., 1971. Pulsation theory of ore formation as shown on several postmagmatic ore deposits in Bulgaria. Geol. Ore Dep., 13(6), 44-56.

Konovalov, I. V., 1073, Wallrock zonation of a gold deposit (Lena Region). Internat. Geol. Rev., 15, 1180-1181.

Konstantinov, M. m., 1974. Structures of gold fields in Transcaucasia and fundamental relationships in their positions. Internal. Geol. Rev., 16, 269-273.

Koo, J. and Mossman, D. J., 1975. Origin and metamorphism of the Flin Flon stratabound Cu-Zn sulfide deposit, Saskatchewan and Maiitoba. Econ. Geol., 70, 48-62.

Koppel, V. H., and Saager, R., 1974. Lead isotipe evidence on the detrital origin of Witwatersrand pyrites and its bearing on the provenance of the Witwatersrand gold. Econ. Geol., 69, 318-331.

Koschmann, A. H., and Bergendahl, M. H., 1968. Principal Gold Producing Districts of the United States. U. S. Geological Survey, Washington, 283 pp.

Koshlakov, I. P., 1969. Age relations between quartz carbonate veins within some intrusions in southwestern Karamazar. Geol. Ore Dep., 11(5), 114-120.

Krauskopf, K. B., 1967. Introduction to Geochemistry. MeGraw-Hill, New York, 721 pp.

Krendelev, F. P., 1970. Sources in gold in Pre-Cambrian conglomerates of sulphide type. In: Pouba and Stemprok (Editors) (q. v.), pp. 113-115.

Kroner, A., 1975. Proterozoic crustal evolution in parts of southern Africa and evidence for extensive sialic crust since the end of the Archean. Mtg. Roy. Soc. Lond., Mar. 13-14.

Kushnarov, I. P., 1969. Depths of Formation of Endogenic Ore Deposits. Nyedra, Moscow, 155 pp.

Kutina, J., 1975. Tectonic development and metallogeny of Madagascar with reference to the fracture pattern of the Indian Ocean. Geol. Soc. Amer. Bull., 86, 582-592.

Lambert, I. B., and Wyllie, P. J., 1970. Low-velocity zone of the Earth's mantle: incipient melting caused by water. Science, 196, 764-766.

Lamey, C. A., 1966. Metallic and Industrial Mineral Deposits. McGraw-Hill, New York, 567 pp.

Laravie, J. A., 1975. Geometry and lateral strain of subducted plates in island arcs. Geology, 3, 484-486.

Lee, T., and Yao, C-L., 1970. Abundance of chemical elements in the Earth's crust and its major tectonic units. Internat. Geol. Rev., 12, 778-786.

Leitch, E. C., 1975. Plate tectonic interpretation of the Paleozoic history of the New England fold belt. Geol. Soc. Amer. Bull,, 86, 141-144.

Le Pichon, X., Francheteau, J., and Bonnin, J., 1973. Plate Tectonics. Elsevier, Amsterdan, 300 pp.

Levinson, A. A., 1974. Introduction to Exploration Geochemistry. Applied Publishing, Calgary, 612 pp.

Li, L. V., Mikheev, V. G., and Rukosuev, M. N., 1971. Probable formation temperature of arsenopyrites of gold deposits in the Yenisei ridge according to the data from X-ray investigations. Geol. Ore Deps, 13(4), 108-112.

Linares, E., and Valencio, D. A., 1975. Paleomagnetism of K-Ar ages of some trachybasaltic dikes from Rio de los Molinos, Province of Cordoba, Republic of Argentina. J. Geophys. Res., 80, 3315-3321.

Lindgren, W., 1911. The Tertiary gravels of the Sierra Nevada of California. U. S. Geol. Surv. Ppr 73, 226 pp.

Lindgren, W., 1933. Mineral Deposits, 4th ed. McGraw-Hill, New York, 930 pp.

Lipman, P. W., and Bunker, C. M., 1972. Upper Cenozoic basalts of the Rocky Mountain region: Geological setting and general petrology. Geol. Soc. Amer. Abstr. Prog., 4, 190-191.

Litvinov, V. L., and Lyakhov, Yu. V., and Popivnyak, I. V., 1970. Paleotemperature zoning of the Kariyskoye gold deposit. Geol. Ore Dep., 12(5), 96-101.

Lord, C. S., 1942. Gunnar gold mine, Manitoba. In: Newhouse (Editor) (q. v.), pp. 253-254.

Loskutov, V. F., and Grebenchikov, A. M., 1974. Structural characteristics of Arkharly gold gield in southwestern Dzhungaria. Internat. Geol. Rev., 16, 400-404.

Lubimova, E. A., and Nikitina, I. N., 1975. On heat flow singularities over mid-ocean ridges. J. Geophys. Res., 80, 232-243.

Macdonald, K., Luyendyk, B. P., Mudie, J. D., and Spiess, F. N., 1975. Near-bottom geophysical study of the Mid-Atlantic Ridge median volley near lat 37°N: Preliminary observations. Geology, 3, 211-215.

Mammerickx, J., Anderson, R. N., Menard, H. W., and Smith, S. M., 1975. Morphology and tectonic evolution of the east-central Pacific. Geol. Soc. Amer. Bull., 86, 111-118.

Marlow, M. S., Scholl, D. W., and Garrison, L. E., 1973. Comparative geologic histories of the Aleutians and Lesser Antilles island arcs. Geol. Soc. Amer. Abstr. Prog., 5, 77-78.

Mason, B., 1966. Principles of Geochemistry, 3rd ed. Wiley, New York, 329 pp.

Masse, R. P., 1975. Baltic Shield crustal velocity distribution. Bull. Seis. Soc. Amer., 65, 885-897.

Mattison, J. M., 1975. Early Paleozoic ophiolite complexes of Newfoundland: Isotopic ages of zircons. Geology, 3, 181-183.

Maucher, A., 1972. Time and stratabound ore deposits and the evolution of the Earth. Internat. Geol. Cong. Abstr., 24, 137.

Mavko, G., and Nur, A., 1975. Melt squirt in the asthenosphere.

J. Geophys. Res., 80, 1444-1448.

Max, M. D., and Riddihough, R. P., 1975. Continuation of the Highland Boundary fault in Ireland. Geology, 3, 206-210.

McElhinny, M. W., 1972. Paleomagnetism and plate tectonics. Internat. Geol. Cong. Abstr., 24, 92.

McGetchin, T. R., Nikhanj, Y. S., and Chodos, A. A., 1973. Carbonate-kimberlite relations in the Cane Valley diatreme, San Juan County, Utah. J. Geophys. Res., 78, 1854-1869.

McGlynn, J. C., and Irving, E., 1974. Paleomagnetism of early Aphebian dibase dykes from the Slave Structural Province, Canada. Tectonophys., 26, 23-38.

McGregor, B. A., and Rona, P. A., 1975. Crest of the Mid-Atlantic Ridge at 26°N. J. Geophys. Res., 80, 3307-3314.

McKenzie, D. P., 1972. Plate tectonics. In: Robertson (Editor), (q. v.), pp. 323-360.

McKinstry, H. E., 1942a. Ballarat, Victoria, Australia. In: Newhouse (Editor) (q. v.), pp. 162-163.

McKinstry, H. E., 1942b. Bendigo, Victoria, Australia. In: Newhouse (Editor) (q. v.), pp. 160-162.

McKinstry, H. E., 1942c. Woods Point, Victoria, Australia. In: Newhouse (Editor) (q. v.), Op. 252-253.

Mercier, J-C, and Carter, N. L., 1975. Pyroxene geotherms. J. Geophys. Res., 80, 3349-3362.

Merwin, R. W., 1968. Gold resources in the Tertiary gravels of California. U. S. Bu. Min. Tech. Prog. Rpt., 14 pp.

Mikhalevskiy, I. V., 1969. The screening effect of porphyries in the lead-zinc deposit of Putintsevskoye. Geol. Ore Dep., 11(5), 98-103.

Milici, R. C., 1975. Structural patterns in the southern Applachians: Evidence for a gravity slide mechanism for Alleghanian deformation. Geol. Soc. Amer. Bull., 86, 1316-1320.

Misik, M. (Editor), 1968. Orogenic Belts. Academia, Prague, 327 pp.

Miyashiro, A., 1972. Metamorphism and related magmatism in plate tectonics. Amer. J. Sci., 272, 656.

Miyashiro, A., 1973a. Metamorphism nd Metamorphic Belts. Halsted, New York, 492 pp.

Miyashiro, A., 1973b. Paired and unpaired metamorphic belts. Tectonophys., 17, 241-254.

Miyashiro, A., 1973c. Paired and unpaired metamorphic belts. Trans. Amer. Geophys. Union, 54, 211.

Molnar, P., and Atwater, T., 1973. On the relative motion of "hotspots". Nature, 246, 288-290.

Moiseenko, V. G., Nechkin, G. S., and Fatyanov, I. I., 1970. Geochemical conditions of redistribution of gold associated with magmatic rocks. In: Pouba and Stemprok (Editors) (q. v.), pp. 115-120.

Moore, E. S., 1940. Genetic relations of gold deposits and igneous rocks in the Canadian shield. Econ. Geol., 35, 127-139.

Moore, J. C., 1975. Selective subduction. Geology, 3, 530-532.

Moore, G. W., and Silver, E. A., 1968. Gold distribution on the sea floor off the Klamath mountains, California. U. S. Geol. Surv. Circ. 605, 9pp.

Moravek, P., and Rus, V., 1970. Regeneration of gold in the central Bohemian pluton area, Czechoslovakia. In: Pouba and Stemprok (Editors) (q. v.), pp. 120-125.

Muehlberger, W. R., and Ritchie, A. W., 1975. Caribbean-Americas plate boundary in Guatemala and southern Mexico as seen on Skylab IV orbital photography. Geology, 3, 232-235.

Mueller, S. (Editor), 1974. Structure of the Earth's Crust Based on Seismic Data. Elsevier, Amsterdam, 391 pp.

Murray, C. G., 1970. Magma genesis and heat flow: Differences between mid-oceanic ridges and African rift valleys. Earth Planet. Sci. Ltr., 9, 34-38.

Naumenko, V. V., and Skarzhinskiy, V. I., 1970. The endogenetic ore formation of Neogene age in Transcarpathia. Geol. Ore Dep., 12(1), 79-82.

Nayborodin, V. I., and Goncharov, V. I., 1970. Temperature during the formation of Agatovskoye gold-silver deposit. Geol. Ore Dep., 12(6), 46-51.

Nekrasova, A. N., 1972. Specific characteristics of mineral composition of ores at the Karamken gold-silver deposit. Geol. Rudn. Mest., 14(3), 45-54.

Newhouse, W. H. (Editor), 1942. Ore Deposits as Related to Structural Features. Princeton Univ. Press, Princeton, 280 pp.

Nitsan, U., 1973. Viscous heat production in a slab. J. Geophys. Res., 78, 1395-1397.

Noble, J. A., 1950. Ore mineralization in the Homestake gold mine, Lead, South Dakota. Geol. Soc. Amer. Bull., 61, 221-252.

Noble, J. A., 1974. Metal provinces and metal finding in western United States. Min. Dep., 9, 1-25.

Noble, J. A., Harder, J. A., and Slaughter, A. L., 1949. Structure of a part of the northern Black Hills and the Homestake mine, Lead, South Dakota. Geol. Soc. Amer. Bull., 60, 321-352.

Norton, J. J., 1974. Gold in the Black Hills, South Dakota, and how new deposits might be found. U. S. Geol. Surv. Circ. 699, 22 pp.

Odman, O. H., 1942. Boliden, Sweden. In: Newhouse (Editor) (q. v.), pp. 166-169.

Oldenburg, D. W., and Brune, J. N., 1975. An explanation for the orthogonality of ocean ridges and transform faults. J. Geophys.

O'Neil, J. R., and Silberman, M. L., 1974. Stable isotope relations in epithermal Au-Ag deposits. Econ. Geol., 69, 902-909.

Ostrovskiy, E. Ya., Portnov, A. M., and Drabkin, I. E., 1970. Aerial survey of gamma ray radiation in prospecting for near-surface gold deposits. Geol. Ore Dep., 12(3), 87-91.

Oversby, V. M., 1975. Lead isotopic systematics and ages of Archean acid intrusives in the Kalgoorlie-Norseman area, Western Australia. Geochim. Cosmochim. Acta, 39, 1107-1125.

Oyarzun, M. J., 1975. Porphyry copper and tin-bearing porphyries: A discussion of genetic models. Phys. Earth Planet. Int., 9, 259-263.

Ozerova, N. A., Aydinyan, N. Kh., Dobrovolskaya, A. F., Zubov, V. I., and Laputina, I. P., 1969. Recent mercury mineralization of the Mendeleyev volcano (Kurile Islands). Geol. Ore Dep., 11(5), 17-33.

Page, R. W., and McDougall, I., 1972. Ages of mineralization of gold and porphyry copper deposits in the New Guinea highlands. Econ. Geol., 67, 1034-1048.

Panov, B. S., and Nikolskiy, I. L., 1972. Mineralogy of the Ostrobugorsklye gold ore deposit in Donets basin. Geol. Rudn. Mest., 14(1), 119-124.

Park, C. F., Jr., 1942. Some gold deposits in Georgia. In: Newhouse (Editor) (q. v.), pp. 199-201.

Park, C. F., Jr., and MacDiarmid, R. A., 1970. Ore Deposits, 2d ed. Freeman, San Francisco, 522 pp.

Petrascheck, W. E. (Editor), 1973. Metallogenic and Geochemical Provinces. Springer, Vienna, 183 pp.

Petrovskaya, N. V., Novikov, V. M., Formin P. S., Frolova, K. E., and Evstignyeva, T. L., 1971. Specific characteristics of native gold as the indicator of subsurface Tertiary deposits in the lower Amur region. Geol. Ore Dep., 13(1), 17-27.

Pettijohn, F. J., 1972. The Archean of the Canadian Shield: A resume. In: Doe and Smith (Editors) (q. v.), pp. 131-149.

Piper, J. D. A., 1975a. Paleomagnetic correlations of Precambrian formations of east-central Africa and their tectonic implications. Tectonophys., 26, 135-161.

Piper, J. D. A., 1975b. Paleomagnetic evidence for a Proterozoic super-continent. Mtg. Roy. Soc. Lond., Mar. 13-14.

Poltorikin, P. I., 1974. Gold contents of pyrites in the Kalba gold belt. Internat. Geol. Rev., 16, 154-155.

Pouba, Z., and Stemprok, M. (Editors), 1970. Problems of Hydrothermal Ore Deposition. Schweitzerbart'sche Verlag., Stuttgart, 396 pp.

Radkevich, E. A., 1971. Metallogenic zones of the Pacific ore belt. Geol. Ore Dep., 13(6), 3-13.

Radkevich, E. A., 1972. Metallogenic zonality of the Pacific ore belt. Internat. Geol. Cong. Abstr., 24, 144.

Ramberg, I. B., and Smithson, S. B., 1975. Gridded fault patterns in a late Cenozoic and Paleozoic continental rift. Geology, 3, 201-205.

Ramsay, W. R. H., and Kobe, H. W., 1974. Great Barrier Island silver-gold deposits, Hauraki Province, New Zealand. Min. Dep., 9, 143-153.

Randall, J. A., 1972. Metallization sequence in the Tayoltita region, San Dimas, Durango, Mexico. In: F. A. Campbell and H. D. B. Wilson (Editors), Mineral Deposits. 24th International Geol. Cong., Ottawa, pp. 309-317.

Razin, L. V., and Yurkina, K. V., 1971. Composition of gold mineralization in disseminated ores of the Norilsk deposit. Geol. Ore Dep., 13(1), 88-92.

Reed, J. C., 1942. Veins in the Warren District, Idaho. In: Newhouse (Editor), (q. v.), p. 175.

Reyment, R. A., 1973. Cretaceous history of the South Atlantic Ocean. In: Tarling and Runcorn (Editors) (q. v.), vol. 2, pp. 805-814.

Richter, F. M., and Parsons, B., 1975. On the interaction of two scales of convection in the mantle. J. Geophys. Res., 80, 2529-2541.

Rickard, D. T., and Zweifel, H., 1975. Genesis of Precambrian sulfide ores, Skellefte district, Sweden. Econ. Geol., 70, 255-274.

Ridge, J. D. (Editor), 1968. Ore Deposits of the United States, 1933-1967. Amer. Inst. Min. Metall. Petrol. Engr., 2 vols.

Riding, R., 1974. Model of the Hercynian foldbelt. Earth Planet. Sci. Ltr., 24, 125-135.

Ringwood, A. E., 1974. The petrological evolution of island arc systems. J. Geol. Soc. Lond., 130, 183-204.

Roaldset, E., 1975. Rare earth element distributions in some Precambrian rocks and their phyllosilicates, Numedal, Norway. Geochim. Cosmochim. Acta, 39, 455-469.

Roberts, R. J., 1966. Metallogenic provinces and mineral belts in Nevada. Nev. Bu. Min. Rpt. 13-A, 47-72.

Roberts, R. G., 1975. The geological setting of the Mattagami Lake mine, Quebec: A volcanogenic massive sulfide deposit. Econ. Geol., 70, 115-129.

Roberts, R. J., Radtke, A. S., and Coats, R. R., 1971. Gold-bearing deposits in north-central Nevada and southwest Idaho. Econ. Geol., 66, 14-33.

Robertson, E. C. (Editor), 1972. The Nature of the Solid Earth. McGraw-Hill, New York, 677 pp.

Rogers, J. J. W., and 7 others, 1974. Paleozoic and lower Mesozoic volcanism and continental growth in the western United States. Geol. Soc. Amer. Bull., 85, 1913-1924.

Rona, P. A., and Dietz, R. S., 1974. Sediment accumulation, sea-floor spreading, and eustacy: Reply. Geol. Soc. Amer. Bull., 85, 831-832.

Rosendahl, B. R., Moberly, R., Halunen, A. J., Rose, J. C., and Kroenke, L. W., 1975. Geological and geophysicsl studies of the Canton Trough region. J. Geophys. Res., 80, 2565-2574.

Rowlands, N. J., 1974. The gitology of some Adelaidean strati-form copper occurrences. In: Bartholome (Editor) (q. v.), pp. 419-427.

Roy, R. F., Blackwell, D. D., and Decker, E. R., 1972. Continental heat flow. In: Robertson (Editor) (q. v.), pp. 506-543.

Rozhkov, I. S., 1968. The characteristic features of gold deposits of mobile belts and platforms. In: Stemprok (Editor) (q. v.), pp. 313-322.

Rozhkov, I. S., 1970. The conditions of origin of auriferous deposits of gold-telluride type. In: Pouba and Strmprok (Editors) (q. v.), pp. 126-129.

Rozhkov, I. S., 1971a. Near-surface deposit of gold. Geol. Ore Dep., 13(3), 3-14.

Rozhkov, I. S., 1971b. Gold deposits of medium depth. Geol. Ore Dep., 13(5), 22-37.

Rye, R. O,, and Ohmoto, H., 1974. Sulfur and carbon isotopes and ore genesis: A review. Econ. Geol., 69, 826-842.

Rye, D. M., and Rye, R. O., 1974. Homestake gold mine, South Dakota: I, Stable isotope studies. Econ. Geol., 69, 293-317.

Rye, D. M., Doe, B. R., and Delevaux, M. H., 1974. Homestake gold mine, South Dakota: II, Lead isotope, mineralization ages, and source of lead of the northern Black Hills. Econ. Geol., 69, 814-822.

Saager, R., and Sinclair, A. J., 1974. Factor analysis of stream sediment geochemical data from the Mount Nansen area, Yukon Territory, Canada. Min. Dep., 9, 243-252.

Sacks, I. S., and Okada, H., 1975. A comparison of the anelasticity structure beneath western South America and Japan. Phys. Earth Planet. Int., 9, 211-219.

Safonov, Yu. G., 1971. The role of block tectonics during the formation of the Lebedinoya gold ore field in central Aldan. Geol. Ore Dep., 13(5), 76-86.

Sakharova, M. S., 1970. On the distribution with depth of sulfosalts and gold in a hydrothermal ore deposit. In: Pouba and Stemprok (Editors) (q. v.), pp. 129-134.

Sakharova, M. S., and Krivitskaya, N. N., 1970. The mineralogical and geochemical characteristics of the lead-cinnabar-bismuth sulfosalts from the gold deposits of the eastern Transbaikal region. Geol. Ore Dep., 12(4), 56-70.

Swakins, F. J., and Rye, D. M., 1974. Relationship of Homestake-type gold deposits to iron-rich Precambrian sedimentary rocks. Trans. Inst. Min. Metall., 83(810), B56-B59.

Schau, M., 1975. Meta-komatiites in a stable Archean environment. Geol. Soc. Amer. Abstr. Prog., 7, 849.

Scheidegger, A. E., 1972. The rheology of the tectonosphere. Internat. Geol. Cong. Abstr., 24, 99.

Schidlowski, M., 1968. The gold fraction of the Witwatersrand conglomerates from the Orange Free State Goldfield (South Africa). Min. Dep., 3, 344-363.

Schilling, J. G., 1975a. Azores mantle blob: Rare-earth evidence. Earth Planet. Sci. Ltr., 25, 103-115.

Schilling, J. G., 1975b. Rare-earth variations across "normal segments" of the Reykjanes Ridge, 60° - 53°N, Mid-Atlantic Ridge, 29°S, and East Pacific Rise, 2° - 19°S, and evidence on the composition of the underlying low-velocity layer. J. Geophys. Res., 80, 1459-1473.

Schilling, J. G., and Bonatti, E., 1975. East Pacific Ridge (2°S - 19°S) versus Nazca interplate volcanism: Rare-earth evidence. Earth Planet. Sci. Ltr., 25, 93-102.

Schonwandt, H. K. V., 1974. Gold from Flat. Norsk Geologisk Tidsskrifft, 54, 63-67.

Schwartz, G. M., 1944. Host minerals of native gold. Econ. Geol., 39, 371-411.

Schwarzer, R. R., and Rogers, J. W., 1973. Alkali olivine basalt series of the Davis Mountains, South Central Texas: A comparison with worldwide occurrences. Geol. Soc. Amer. Abstr. Prog., 5, 102.

Schwarzer, R. R., and Rogers, J. J. W., 1974. A worldwide comparison of alkali olivine basalts and their differentiation trends. Earth Planet. Sci. Ltr., 23, 2860296,

Sclater, J. G., and Fisher, R. E., 1974. Evolution of the East Central Indian Ocean, with emphasis on the tectonic setting of the Ninetyeast Ridge. Geol. Soc. Amer. Bull., 85, 683-702.

Seeger, C. R., 1975. Absolute motion of the Eurasian plate: A problem in vector geometry: Comment. Geology, 3, 159.

Sestini, G., 1973. Sedimentology of a paleoplacer: The gold-bearing Tarkwaian of Ghana. In: Amstutz and Bernard (Editors) (q. v.), pp. 275-305.

Shackleton, R. M., 1975. Pan African structures. Mtg. Roy. Soc. Lond., Mar. 13-14.

Shcherbakov, Yu. G., 1970. Source of gold in endogenous deposits. In: Pouba and Stemprok (Editors),(q. v.), pp. 134-136.

Shenon, P. J., and Reed, J. C., 1942. Veins in the Elk City district, Idaho. In: Newhouse (Editor) (q. v.), pp. 175-176.

Shoemaker, E. M. (Editor), 1975. Continental Drilling. Carnegie Institution, Washington, 56 pp.

Shpunt, B. R., 1970. Platinum group minerals in Quaternary sediments of the Anabar-Olenek interfluve. Geol. Ore Dep., 12(2), 123-126.

Shtreys, N. A., and Tseysler, V. M., 1972. Formation of geosynclinal regions. Geotectonics, 1, 1-2.

Silberman, M. L., Chesterman, C. W., Kleinhampl, F. J., and Gray, C. H., Jr., 1972. K-Ar ages of volcanic rocks and gold-bearing quartz-adularia veins in the Bodi Mining district, Mono County, California. Econ. Geol., 67, 597-604.

Silberman, M. L., Berger, B. R., and Koski, R. A., 1974. K-Ar age relations of granodiorite emplacement and tungsten and gold mineralization near the Getchell mine, Humboldt County, Nevada. Econ. Geol., 69, 646-656.

Silichev, M. K., 1970. The geological position and structural features of the Nezhdaninskoye gold deposit. Geol. Ore Dep., 12(2), 96-102.

Sillitoe, R. H., 1972a. A plate tectonic model for the origin of porphyry copper deposits. Econ. Geol., 67, 184-197.

Sillitoe, R. H., 1972b. Relation of metal provinces in western America to subduction of oceanic lithosphere. Geol. Soc. Amer. Bull., 83, 813-818.

Sillitoe, R. H., 1972c. Formation of certain massive sulphide deposits at sites of sea-floor spreading. Trans. Inst. Min. Metall., 81, B141-B148.

Sillitoe, R. H., 1973. Geology of the Los Palambres porphyry copper deposit, Chile. Econ. Geol., 68, 1-10.

Sipkin, S. A. and Jordan, T. H., 1975. Lateral heterogeneity of the upper mantle determined from the travel times of ScS. J. Geophys. Res., 80, 1474-1484.

Slaughter, A. L., 1968. The Homestake mine. In: Ridge (Editor) (q. v.), vol. 2, pp. 1436-1459.

Spall, H,, 1972. Paleomagnetism and Precambrian continental drift. Internat. Geol. Cong. Abstr., 24, 101.

Spence, C. D., 1975. Volcanogenic features of the Vauze sulfide deposit, Noranda, Quebec. Econ. Geol., 70, 102-114.

Spencer, A. M. (Editor), 1974. Mesozoic-Cenozoic Orogenic Belts: Data for Orogenic Studies. Scottish Academic, Edinburgh, 809 pp.

Spooner, J., Williams, L., Kennedy, A., and Burcher, M., 1974. Mining Annual Review. Mining Journal, London, 506 pp.

Steinhorn, T. L., 1973. Depth-of-origin of magmas of the high cascades as inferred from geochemical relations. Geol. Soc. Amer. Abstr. Prog., 5, 197.

Stemprok, M. (Editor), 1968. Endogenous Ore Deposits. Academia, Prague, 425 pp.

Stephansson, O., 1975. Polydiapirism of granitic rocks in the Svecofennian of central Sweden. Precamb. Res., 2, 189-214.

Stevenson, J. S., 1942. Gold-quartz veins, O. K. Mountain, Rossland, British Columbia. In: Newhouse (Editor) (q. v.), pp. 246-247.

Stewart, J. H., and Poole, F. G., 1975. Extension of the Cordilleran miogeosynclinal belt to the San Andreas fault, Southern California. Geol. Soc. Amer. Bull., 86, 205-212.

Stewart, J. H., Walker, G. W., and Kleinhampl, F. J., 1975. Oregon-Nevada lineament. Geology, 3, 265-268.

Strong, D. F., 1973. Volcanic, intrusive, and teconic activity in an Ordovician island arc of central Newfoundland. Geol. Soc. Amer. Abstr. Prog., 5, 224.

Strong, D. F., and Stewart, R. K., 1974. Contrasting Archean and Proterozoic geological regimes: A possible thermal explanation. Nature, 249, 545.

Subbothin, S. I., Sollogub, V. B., Khain, V. Y., Slavin, V. I., and Chekunov, A. V., 1972. Some remarks concerning the structure and evolution of the Earth's crust. Hung. Geophys. Inst. Geophys. Trans., 1972, 151-165.

Sutton, J., Shackleton, R. M., and Briden, J. C., 1975. Global tectonics in Proterozoic times. Mtg. Roy. Soc. Lond., Mar. 13-14.

Sutulov, A., 1973. Mineral Resources and the Economy of the USSR. McGraw-Hill, New York, 192 pp.

Symons, D. T. A., 1975. Huronian glaciation and polar wander from the Gowganda Formation, Ontario. Geology, 3, 303-306.

Tandon, A. N., and Srivastava, H. N., 1975. Focal mechanism of some recent Himalayan earthquakes and regional plate tectonics. Bull. Seis. Soc. Amer., 65, 963-969.

Tarling, D. H., 1973. Metallic ore deposits and continental drift. Nature, 243, 193-196.

Tarling, D. H., and Runcorn, S. K. (Editors), 1973. Implica-

tions of Continental Drift to the Earth Sciences, 2 vols. Academic, New York, 1184 pp.

Tatsch, J. H., 1959. Initial results of preliminary investigations (1938-1958) regarding the feasibility of using polyistic planetary configurations for making certain solar-system evolutionary studies. U. S. Embassy, Caracas, Venezuela, in manus.

Tatsch, J. H., 1960. The evolution of the solar system in accordance with a dual primeval planet model. U. S. Embassy, Caracas, Venezuela, private publication, given limited distribution to approximately 300 slected Earth and planetary scientists in various parts of the world.

Tatsch, J. H., 1962. The evolution of the retrograde motion of certain planetary satellies in accordance with a dual primeval planet model, outline of doctoral dissertation, submitted to graduate advisor (the late Gerard P. Kuiper), February 4, 1962, about 75 pages.

Tatsch, J. H., 1963a. Certain selenophysical implications of applying a dual primeval planet model to the Earth. _Trans. Amer. Geophys. Union_, _44_, 877.

Tatsch, J. H., 1963b. Certain seismological implications of applying a dual primeval planet model to the Earth. _Trans. Amer. Geophys. Union_, _44_, 887.

Tatsch, J. H., 1963c. Certain volcanological implications of applying a dual primeval planet model to the Earth. _Trans. Amer. Geophys. Union_, _44_, 892.

Tatsch, J. H., 1964a. Distribution of active volcanoes: Summary of preliminary results of three-dimensional least-squares analysis. _Geol. Soc. Amer. Bull._, _75_, 751-752.

Tatsch, J. H., 1964b. Certain geomagnetic implications of applying a dual primeval planet model to the Earth. _Trans. Amer. Geophys. Union_, _45_, 39.

Tatsch, J. H., 1964c. Certain oceanographical implications of applying a dual primeval planet model to the Earth. *Trans. Amer. Geophys. Union*, *45*, 74.

Tatsch, J. H., 1964d. Further volcanological implications of applying a dual primeval planet model to the Earth. *Trans. Amer. Geophys. Union*, *45*, 124.

Tatsch, J. H., 1965a. The global distribution patterns of fossil volcanoes and their relationship to the global distribution patterns of active volcanoes. National Science Foundation proposal number P51073R.

Tatsch, J. H., 1966a. Certain correlations between geomagnetic observations in the Indian Ocean and deductions arrived at from applying a dual primeval planet model to that region of the Earth. *Trans. Amer. Geophys Union*, *47*, 56-57.

Tatsch, J. H., 1966b. Certain correlations between oceanographic observations along the South Atlantic Ridge and deductions arrived at from applying a dual primeval planet model to that region of the Earth. *Trans. Amer. Geophys. Union*, *47*, 123.

Tatsch, J. H., 1966c. Certain correlations between geomagnetic observations in the South Atlantic Ocean and deductions arrived at from applying a dual primeval planet model to that region of the Earth. *Trans. Amer. Geophys. Union*, *47*, 464.

Tatsch, J. H., 1966d. Certain correlations between oceanographic observations along the North Atlantic Ridge and deductions made from applying a dual primeval planet model to that region of the Earth. *Trans. Amer. Geophys. Union*, *47*, 477.

Tatsch, J. H., 1966e. Certain correlations between selenophysical observations and deductions arrived at from applying a dual primeval planet model to the Earth-Moon system. *Trans. Amer. Geophys. Union*, *47*, 486.

Tatsch, J. H., 1966f. Certain correlations between seismological observations in the African continent and deductions made from applying a dual primeval planet model to that region of the

Earth. Trans. Amer. Geophys. Union, 47, 490.

Tatsch, J. H., 1967. Global geomagnetic evidence supporting the existence of a geophysical equator predicted as a consequence of applying a dual primeval planet model to the Earth. Trans. Amer. Geophys. Union, 48, 59.

Tatsch, J. H., 1969. Sea-floor spreading, continental drift, and plate tectonics unified into a single global concept by the application of a dual primeval planet hypothesis to the Earth. Trans. Amer. Geophys. Union, 50, 672.

Tatsch, J. H., 1970. Global seismicity patterns as interpreted in accordance with a dual primeval planet hypothesis. Geol. Soc. Amer. Abstr. Prog., 2, 153.

Tatsch, J. H., 1972a. The Earth's Tectonosphere: Its Past Development and Present Behavior. Tatsch Assoc., Sudbury, 889 pp.

Tatsch, J. H., 1972b. Global symmetry of mineralization belts. Geol. Soc. Amer. Abstr. Prog., 4, 729-730.

Tatsch, J. H., 1973a. Mineral Deposits: Origin, Evolution, and Present Characteristics. Tatsch Assoc., Sudbury, 264 pp.

Tatsch, J. H., 1973b. The origin, evolution, and present characteristics of the Tertiary petroleum deposits of the world. Tatsch Assoc., Sudbury, 58 pp.

Tatsch, J. H., 1973c. The origin, evolution, and present characteristics of the Mesozoic petroleum deposits of the world. Tatsch Assoc., Sudbury, 62 pp.

Tatsch, J. H., 1973d. The origin, evolution, and present characteristics of the Paleozoic petroleum deposits of the world. Tatsch Assoc., Sudbury, 63 pp.

Tatsch, J. H., 1973e. The origin, evolution, and present characteristics of the giant oil fields of North America. Tatsch Assoc., Sudbury, 67 pp.

Tatsch, J. H., 1973f. The origin, evolution, and present characteristics of the petroleum deposits in the Arctic Ocean and

North Sea areas. Tatsch Assoc., Sudbury, 62 pp.

Tatsch, J. H., 1973g. The origin, evolution, and present characteristics of the petroleum deposits in the Middle East areas. Tatsch Assoc., Sudbury, 62 pp.

Tatsch, J. H., 1973h. The origin, evolution, and present characteristics of the petroleum deposits within the Indonesian islands and within other areas of the arc extending roughly from the Bay of Bengal to the Kermadec-Tonga region. Tatsch Assoc., Sudbury, 66 pp.

Tatsch, J. H., 1973i. The origin, evolution, and present characteristics of the Tertiary basins that appear to be largely nonpetroliferous. Tatsch Assoc., Sudbury, 62 pp.

Tatsch, J. H., 1973j. The probability for the existence of Tertiary petroleum deposits between Cape Cod and Cape Fear. Tatsch Assoc., Sudbury, 65 pp.

Tatsch, J. H., 1973k. Petroleum in the Philippines: An analysis of related Mesozoic and Cenozoic geophenomena. Tatsch Assoc., Sudbury, 43 pp.

Tatsch, J. H., 1974a. *The Moon: Its Past Development and Present Behavior*. Tatsch Assoc., Sudbury, 338 pp.

Tatsch, J. H., 1974b. *Petroleum Deposits: Origin, Evolution, and Present Characteristics*. Tatsch Assoc., Sudbury, 378 pp.

Tatsch, J. H., 1974c. Reversals of the Earth's magnetic field as interpreted by the Tectonospheric Earth Model. Nagata Conference, June 3-4, Pittsburgh, Penna.

Tatsch, J. H., 1975a. *Copper Deposits: Origin, Evolution, and Present Characteristics*. Tatsch Assoc., Sudbury, 339 pp.

Tatsch, J. H., 1975b. *The Physics of the Solar System: An Analysis Based on the Dual Primeval Planet Hypothesis*. Tatsch Assoc. Sudbury, about 450 pp, in prep.

Tatsch, J. H., 1975c. *The Principle of Universal Heterogeneity: The Basis for the Dual Primeval Planet Hypothesis*. Tatsch Assoc.,

Sudbury, about 475 pp., in prep.

Tatsch, J. H., 1975d. The Earth's Tectonosphere: Its Past Development and Present Behavior, 2nd ed. Tatsch Assoc., Sudbury, about 550 pp., in press.

Tatsch, J. H., 1975e. Geothermal Deposits: Origin, Evolution, and Present Characteristics. Tatsch Assoc., Sudbury, about 340 pp., in press.

Tatsch, J. H., 1975 f. Uranium Deposits: Origin, Evolution, and Present Characteristics. Tatsch Assoc., Sudbury, about 320 pp., in press.

Tatsch, J. H., and Talbot, J. L., 1971. Global symmetry of tectonic belts. Geol. Soc. Amer. Abstr. Prog., 3, 728-729.

Taylor, H. P., 1973. O^{18}/O^{16} evidence for meteoric-hydrothermal alteration and ore deposition in the Tonopah, Comstock Lode, and Goldfield mining districts, Nevada. Econ. Geol., 68, 747-764.

Taylor, P. N., 1975. An early Precambrian age for migmatitic gneisses from Vilon i Bo, Vesteralen, North Norway. Earth Planet. Sci. Ltr., 27, 35-42.

Terentiev, V. M., Belyayev, G. M., and Kudrayavtsev, B. E., 1971. Metallogenic characteristics of quartz-feldspar metasomatites of the Aldan Shield. Geol. Ore Dep., 13(3), 41-50.

Terman, M. J., 1974. Pre-Mesozoic plate tectonics of the Far East. Geol. Soc. Amer. Abstr. Prog., 6, 982.

Thomas, D. E., 1953. Mineralization and its relationship to the geological structure of Victoria. Fifth Empire Min. Metall. Cong., 971-985.

Thurlow, J. G., Swanson, E. A., and Strong, D. F., 1975. Geology and lithogeochemistry of the Buchans polymetallic sulfide deposits, Newfoundfand. Econ. Geol., 70, 130-144.

Tilling, R. I., Gottfried, D., and Rowe, J. J., 1973. Gold abundance in igneous rocks: Bearing on gold mineralization. Econ. Geol., 68, 168-183.

Tomich, S. A., 1974. A new look at Kalgoorlie Golden Mile geology. Proc. Australasian Inst. Min. Metall., 251, 27-35.

Trofimov, V. S., 1972. On the origin of diamondiferous eclogite. Internat. Geol. Cong. Abstr., 24, 153.

Tvalchrelidze, G. A., 1970. The principal metallogenic epochs of our planet. Geol. Ore Dep., 12(1), 22-36.

Van Schmus, W. R., 1975. Middle Precambrian rocks of the Great Lakes area, North America. Mtg. Roy. Soc. Lond., Mar. 13-14.

Vasuden, V. N., and Naganna, C., 1973. Mineralogy of gold-quartz-sulfide reefs of Hutti gold mines, Raichur district, Mysore State. J. Geol. Soc. Ind., 14, 329-341.

Veizer, J., 1975. 87-Sr/86-Sr composition of seawater during geologic history and its significance as an index of crustal evolution. Geol. Soc. Amer. Abstr. Prog. 7, 874.

Vemban, N. A., Subramanian, K. S., and Meenakshi-Sundaram, S., 1972. Gold occurrences near Kotagiri, Nilgiri, district, Tamil Nadu. Ind. Min., 26(2), 37-41.

Vincent, E. A., and Crockett, J. H., 1960. Studies in the geochemistry of gold: I, The distribution of gold in rocks and minerals of the Skagergaard intrusion, East Greenland. Geochim. Cosmochim. Acta, 18, 130-142.

Viswanathan, S., 1974. Basaltic komatiite occurrences in the Kolar gold field of India? Geol. Mag., 111, 353-254.

Vogt, P. R., 1975. Changes in geomagnetic reversal frequency at times of tectonic change: Evidence for coupling between core and upper mantle processes. Earth Planet. Sci. Ltr., 25, 313-321.

Vogt, P. R., and Johnson, G. L.. 1975. Transform faults and longitudinal flow below the Midoceanic ridge. J. Geophys. Res., 80, 1399-1428.

Walker, J. F., 1942. Gold-quartz veins of the Sheep Creek camp, British Columbia. In: Newhouse (Editor) (q. v.), pp. 177-178.

Wanless, R. K., Boyle, R. W., and Lowden, J. A., 1960. Sulfur isotope investigations of the gold-quartz deposits of the Yellowknife district. Econ. Geol., 55, 1591-1621.

Watanabe, M., 1942a. Epithermal deposits in Japan. In: Newhouse (Editor) (q. v.), pp. 236-238.

Watanabe, M., 1942b. Contact metasomatic deposits in Japan and Korea. In: Newhouse (Editor) (q. v.), pp. 263-164.

Watson, J. V., 1975. Vetical movements in Proterozoic structural provinces. Mtg. Roy. Soc. Lond., Mar. 13-14.

Watterson, J., 1975. Proterozoic deformation patterns in Greenland. Mtg. Roy. Soc. Lond., Mar. 20-21.

Watts, A. B., and Talwani, M., 1975. Gravity effect of downgoing lithospheric slabs beneath island arcs. Geol. Soc. Amer. Bull., 86, 1-4.

Weber, W., and Stephenson, J. F., 1973. The content of mercury and gold in some Archean rocks of hhe Rice Lake (Manitoba) area. Econ. Geol., 68, 401-407.

Weissberg, B. G., 1969. Gold-silver ore-grade precipitates from New Zealand thermal waters. Econ. Geol., 64, 95-108.

Wells, J. D., and Mullens, T. E., 1973. Gold-bearing arsenian pyrite determined by microprobe analysis, Cortez and Carlin gold mines, Nevada. Econ. Geol., 68, 187-201.

Wells, J. D., Stoisser, L. R., and Elliott, J. E., 1969. Geology and geochemistry of the Cortez gold deposit, Nevada. Econ. Geol., 64, 526-537.

White, W. H., 1943. Mechanism and environment of gold deposition in veins. Econ. Geol., 38, 512-532.

White, D. S., 1967. Mercury and base-metal deposits with associated thermal and mineral waters. In: Barnes (Editor) (q. v.), pp. 575-631.

Whitehead, W. L., 1942. The Mother Lode system in southern Eldorado and Amador counties, California. In: Newhouse (Editor) (q. v.), pp. 178-182.

Whitehead, J. A.,Jr., and Luther, D. S., 1975. Dynamics of laboratory diapir and plume models. J. Geophys. Res., 80, 705-717.

Whiteman, A., Naylor, D., Pergum, R., and Rees, G., 1975. North Sea troughs and plate tectonics. Tectonophys., 26, 39-54.

Wilson, E. D., 1942. Arizona lode gold deposits. In: Newhouse (Editor) (q. v.), pp. 242-243.

Wilson, R. C. L., 1975. Atlantic opening and Mesozoic continental margin basins of Iberia. Earth Planet. Sci. Ltr., 25, 33-43.

Winterer, E. L., 1973. Sedimentary facies and plate tectonics of equatorial Pacific. Amer. Assoc. Petrol. Geol. Bull., 57, 265-282.

Wise, E. M., 1964. Gold: Recovery, Properties, and Applications. Van Nostrand, Princeton, 367 pp.

Wright, C., 1975. Comments on "Seismic array evidence of a core boundary source for the Hawaiian linear volcanic chain" by E. R. Kanesewich et al. J. Geophys. Res., 80, 1915-1919.

Wright, J. B., and McCurry, P., 1973. Sea-floor spreading and continental ore deposits. In: Tarling and Runcorn (Editors) (q. v.), vol. 1, pp. 563-565.

Wyllie, P. J. (Editor), 1967. Ultramafic and Related Rocks. Wiley, New York, 464 pp.

Wyllie, P. J., 1969a. The origin of ultramafic and ultrabasic rocks. Tectonophy., 7, 437-455.

Wyllie, P. J., 1969b. The ultramafic belts. In: Hart (Editor) (q. v.), pp. 480-488.

Wyllie, P. J., 1971. The Dynamic Earth: Textbook in Geosciences. Wiley, New York, 416 pp.

Wynne-Edwards, H. R. (Editor), 1969. Age Relations in High-grade Metamorphic Terrains. Geol. Soc. Can., Toronto, 228 pp.

Wynne-Edwards, H. R., and Hasan, Z., 1970. Intersecting orogenic belts across the North Atlantic. Amer. J. Sci., 268, 289-308.

Yablokova, S. V., and Ryzhov, B. V., 1973. Role of "old gold" in supplying Quaternary placers of Mariinsk Taiga. Internat. Geol. Rev., 15, 1182-1185.

Yeleyeva, I. V., and Rusinova, O. V., 1969. Genesis of quartz-kaolinite rock alteration in the upper levels of the Baley (gold) deposits and its importance in prospecting. Geol. Ore Dep., 11(4), 88-95.

York, J. E., and Helmberger, D. V., 1973. Low-velocity zone variations in the southwestern United States. J. Geophys. Res., 78, 1883-1886.

Yoshii, T., 1975. Regionality of group volocities of Rayleigh waves in the Pacific and thickening of the plate. Earth Planet. Sci. Ltr., 25, 305-312.

Yudin, I. M., 1971. Finely dispersed sulfide mineralization of surrounding sedimentary-metamorphic series in the Maruntau deposit and some problems of its origin. Geol. Ore Dep., 13(2), 43-51.

Zagruzina, I. A., 1970. The age of gold mineralization of the northeast of the Soviet Union. Geol. Ore Dep., 12(1), 79-82.

Zagruzina, I. A., 1971. The age of the Busugun-Khatynnakl gold-bearing intrusive complex. Geol. Ore Dep., 13(4), 99-102.

Zakharov, Ye., Ye., Novikov, V. P., and Pilipenko, G. N., 1969. Occurrence of gold in Mesozoic metasomatites in the crystalline basement of the central Aldan region. Geol. Ore Dep., 11(2), 85-88.

Zapletal, K., 1968. Tektogene der Erde. In: Mistik (Editor) (q. v.), pp. 213-224.

Zonenshain, L. P., 1972. Similarities in the evolution of geosynclines of different types. Internat. Geol. Cong. Abstr., 24, 106.

Zwart, H. J., 1975. Proterozoic metamorphism in Europe. Mtg. Roy. Soc. Lond., Mar. 13-14.

INDEX

Acadian 53
Accretion, early Earth 19
Active regions 25
Active volcanoes 32
Adelong 118
Adycha 118
Africa 1, 28, 54, 76, 165, 192, 194, 205, 210
African shield 2
Agatovskoye 79
Agua Fria 82
Ajo 82
Alabama 166
Alaska 76, 79, 119
Alaskan lodes 79
Alaskan placers 80
Aldan placers 80
Aldan shield 119
Alder Gulch 209
Aleghany 84
Alelong 79
Aleutians 32
Alpine 53
Alpine peridotites 48
Altai-Sayny 4, 143
Altau 4
Amador 85
American River 80, 85, 209
Americas 1, 50, 194, 205
Amur 3, 80, 119, 209
Anabar 196
Anastra 109
Andes 209
Animas Hills 158
Antiquia 81
Ankerite 203
Antler 87, 212
Apex 101
Appalachian 166
Arabia 196, 210
Arakh 196
Arbacoochee 166
Archean
- deposits 1, 5, 6, 73, 78
- dynamic behavior 194
- enigmas 194

Arctic islands 77
Areas of completed folding 5
Arizona 81
Arizona lodes 167
Ascensive mechanism 52
Ashanti 196
Ashland 128
Asia 1, 4, 118, 194, 205
Atlantic Creek 191
Aurora 82
Australia 1, 3, 47, 75, 82, 118, 165, 192, 194, 205, 211, 212, 216
Baffin Bay 48

Bahia 184
Baker 120
Bald Mountain 143, 186
Balei 5, 120
Baley - see Balei
Balkhash 176
Ballarat 82, 121, 144, 197, 208, 209
Baltic 216
Bangladesh 75
Bankfield 168
Banner 82
Barkerville 144
Basic economics of finding hidden gold deposits 215
Basic 5400-km primordial Earth 10
Basin and Range 83
Bayard 95
Bear 195
Belmoral 203
Bendigo 145, 197, 208
Bengal 32
Beresovsk 4
Bering Sea 76, 104, 211
Beringa 145
Big Bell 145, 189
Big Bug 82
Bilire 196
Bingham 83
Bisbee 82
Blackwood 146
Bland River 2
Bodie 83
Boise Basin 83, 121, 209
Boliden 168
Bolivia 84, 209, 210
Bonanza 89, 146, 215
Bouvet 32
Bradshaw 167
Bralorne 122
Brazil 165, 192, 196, 209, 210, 216
Brewer 122
Britannia 123
Broadlands 139
Brownhill 197, 198, 200, 208
Brunei 75
Buffalo 203
Bulafo 84
Bulawayo 196, 198
Bunker Hill 84
Buolkaloah 196
Burke 146
Burma 75
Busugun 123
Bushveld 3, 169
Byproduct gold 212
Cabarrus 146
Cable 123
Cadillac 169
Calaveras 85
Caldas 81, 84
California 84, 211

Cambrian 4
Camp Floyd 86
Canada 1, 76, 143, 165, 169, 192, 194, 196, 210, 212, 216
Carbonate 159
Carlin 86, 87, 207, 211, 212, 215
Castlemaine 124
Catron 87
Conozoic 1
Central 95
Central Aldan 124, 146
Central America 76
Central Bohemia 147
Central City 196, 198
Central Eureka 88
Ceylon 211
Chalk Creek 89
Champion 198, 201
Charters Towers 136, 147
Chatalko 147
Che-Cheka 137
Cherokee 173
Cherry Creek 167
Chibougamou 170
Chichagoff 79
China 1, 143, 195
Chinkuashih 88
Chukot - see Chukotka
Chukotka 3, 6, 79, 88, 105, 124, 129
Churchill 195
Circum-Pacific 50
Coast Range 80
Cobar 148
Cochrane Hill 124
Cockshutt 180
Clifton 82
Clunes 124
Colfax 170
Colombia 88, 209
Colorado 89, 198
Colorado Plateau 81
Composite-fragment accretionary mode 21
Comstock 91, 208
Cones of activity 55
Congress 167
Connor Creek 120
Consolidated structures 4
Continent-continent collisions 48
Continents 39, 73
Continental drift 29, 40
Continental shields 2, 37
Cook Inlet 80, 119
Coolgardie 171, 189
Copper 50
Cornucopia 120
Cortez 87, 91, 207, 211, 212, 215
Cracker Creek 120
Creede 89, 148
Creighton 173

Cresson 91
Cretaceous placers 215
Cripple Creek 6, 89, 90, 91, 208
Crustal evolution 56
Crustal rifting 39, 75
Cue 189
Dahlonega 171
Damaran 53
Darsun 148, 159
Davidson 148
Deadwood Gulch 171, 186
Deep faults 2
Delamar 92
Delnite 203
Descending lithospheric plates 48
Detour Lake 198, 203
Dharwar 201
Dikdik 171
Disseminated gold in bedrock 211, 215
Distribution of gold deposits 1, 8
Dives 109
Dome 184, 197, 199, 203, 208
Dorn 125
Dos Cabezas 82
Douglas Creek 191
Downieville 84
Driving mechanism for Earth's internal behavior 13, 30 36
Ducktown 188
Dunolly 125
Eagle Creek 120
Earliest gold deposits 1
Earth
evolution of present surficial features 42
geometrical, mechanical, thermal, and chemical behavior 36
internal behavior 30
most probable surficial features 4.6 b. y. ago 40
most probable surficial features 3.6 b. y. ago 41
newest features 27
oldest features 27
seismicity 29
Earthquakes 29, 37
East Belt 84
East China Sea 76
East Dome 203
Eastern USA 209
Eburnian 53
Ecuador 93, 209
Egypt 209
El Dorado 85
El Oro 208
Elk City 93
Elk Creek 171
Elk Mountain 159
Empire 149, 197, 208

Energy sources 56
Enisey - see Yenisei
Eugeosynclines 3, 50
Eurasia 76, 195
Europe 195
Exploration for gold deposits using the Tectonospheric Earth Model as a supplementary tool 10, 45, 214
Extrusive activity 38, 51
Fairbanks 80, 93, 209
Fairfield Valley 149
Fault lines 30
Faunal assemblages 48
Femic minerals 50
Fierro 95
Finland 165
Flåt 172
Flin Flon 199
Folsom 94
Formation of deposits 1
Fortuna 167
Fossil placers 210, 216
Fossil regions 26
Fracture system 25
Franklin 172
Fremont 94
Front Range 90
Galapagos 32
Garden 172, 186
Garden Gulley 126
Gebeit 149
Geoid 38
Geomagnetic dynamo 29
Geomagnetism 39
Georgia 172
Geosynclines 38
Geotectonic structures 1
Geothermal activity 37
Getchell 87, 94, 212
Getygara 4
Ghana 2, 173, 187, 196
Gibraltar 32
Gila City 82
Globe 82
Gold Acres 87, 94, 212
Gold Brick 89
Gold deposits
aluminous-rock associations 207
Archean 194, 205, 216
byproduct from base-metal ores 212
carbontte-rock associations 207
Cenozoic 73, 75, 116
crustal-type associations 206
disseminated in bedrock 211
enigmas within distribution patterns 206
epithermal bonanzas 208
felsic-rock associations 207
fossil placers 210
hot springs associated with 212

Gold Deposits (continued)
marine placers 211
Mesozoic 118, 141
Paleozoic 143, 163
porphyry-copper associations 206
Precambrian 165, 192, 19
194, 205, 216
present distribution as function of Earth's behavior during past 4.6 b. y. 206
Proterozoic 165, 192, 216
rock-type associations 207
spatio-temporal associations 206
siliceous-rock associations 207
sulfides 206
types 206, 214
young placers 2 9
Gold formations 6
Gold Hill 128, 173
Gold-quartz 6
Gold-quartz lodes 207, 216
Gold-quartz sulfides with gold-quartz 7
Gold-quartz sulfides with gold-quartz and gold-silver 8
Gold-quartz with gold-quartz sulfides, and gold-copper 7
Gold-quartz with gold-quartz sulfides in activated areas 8
Gold-quartz with gold-quartz sulfides in geosynclines 7
Gold-quartz with gold sulfides 7
Gold-silver formations 5
Gold-silver with silver-gold 8
Gold sulfides with copper pyrites and polymetallic pyrites 6
Gold sulfides with copper sulfides and platinoids 6
Gold-uranium placers 216
Golden Cycle 91, 94
Golden Drip 134
Golden Mile 173, 175
Goldfield 6, 208
Golima 81, 95
Gore-Tenmile 90
Granada 199
Granite 126
Grant 126
Grant County 95
Grass Valley 84, 150, 208
Gravels 162
Gravity field 38
Gravity tectonics 48
Great Barrier Island 96
Great Fingall 173
Greenhorn 120, 126
Guanajuato 96, 208
Gulf of Aden 46

Gunnar 150
Gunnison 89
Guyana 196, 210, 216
Gwalia 173
Hahns Peak 97
Haile 126, 174
Hallnor 203
Hammonton 97
Hanover 95
Hard Rock 174
Hargreaves 200
Harmonic analyses 35
Harquahala 82, 167
Harricanaw 184
Hartley 196, 199
Haubian 53
Hauraki 97
Heat flow, preferential paths of 29, 37, 58
Hedley 127
Heraclitus 195
Hercynian 53
Hidden Treasure 102
Hill City 174, 186
Hillend 150
Hillsboro 158
Hog Mountain 166, 174
Holkol 127
Hollinger 184, 197, 200, 203 208
Homestake 6, 98, 186, 196, 207, 208, 216
Hughes 200
Hustlers 127
Igneous activity
    description 29
    periodicity in regional and global patterns 206
Igneous rocks 52
Inactive regions 26
Independence 91, 109
India 76, 165, 192, 198, 201, 216
Indian Ocean 46
Indigirka 118
Individual octantal motion of the Earth 15
Indonesian arc 48, 75, 76, 211
Inter-arc basins 48
Intrusive activity 38, 51
Iran 76
Island arcs 47
I. X. L. 134
Idaho 127
Jackson 128
Jacksonville 128
Jacobina 175, 210
Jano-Kolyma 195
Japan 98, 208, 212
Jarbridge 98
Java 75
Jerome 81
Josephine 128
Juneau 79, 98, 197, 208
Jurassic 4, 5

Kainantu 99, 106
Kalgoorlie 175, 196, 198, 200, 216
Kalimantan 75
Kamchatka 6, 100
Karamazar 150
Kariyskoye 129, 151
Katangan 53
Kazakhstan 1, 4, 118, 195
Kennedy 99
Kenoran orogeny 51
Kenterlau 176
Keperveem 129
Kermadecs 32
Kerr-Addison 197, 200, 208
Keystone 99, 176, 186
Khatygyn 196
Khatynnakh 123
Khmer Republic 75
Kilbaran 53
Kilo 196, 200
Kinkwaseki 99
Kirkland Lake 196, 197, 198, 199, 200, 201, 208, 216
Klamath Mountains 130
Klondike 99, 209
Klyuchi 5
Kochkar 4
Kodiak 119
Kofa 81, 167
Kolar 196, 201, 208, 216
Korea 130
Koryako 6, 100
Kolyma 100, 118, 195
Kuraminskaya 147
Kuril 100
Kuznetsk 4
La Grange 100
Lake City 89, 130
Lake Shore 200, 208
Laramide 63
Larder Lake 196, 200, 201, 216
Last Chance Gulch 132
Lead 177
Lebedinoye 130
Leitch 178
Lena 3, 151, 178, 209
Leonora 179
Libby Quadrangle 131
Lincoln 100
Little Long Lac 179
Liturga 119
Loddon Valley 131
Lode deposits 215
Lower Burnt River Valley 120
Low-velocity zone
    description 23
    related to magma 61
Lumpkin 173
Luxemburg 158
Lynn 151
Macassa 200
MacLeod 180
Magmatic activities 1

Magmatic belts - see seismo-
tectonomagmatic belts
Magmatic intrusion 52
Mahoney 100
Maitland 172
Malarctic 199, 201
Malaysia 75, 76, 211
Malheur 131
Maltano 79
Mammouth 167
Manhattan 152
Mantle convection 29
Margadan 88
Mariinsk 100
Marine placers 211
Maslovskiy 152
Matachewan 180
Mayflower 102
McIntyre 184, 197, 202, 203, 208
McWatters 180
Mecklenburg 153
Mediterranean-Indonesian belt 1, 73
Meekatharra 181
Mendeleyev 100
Merced 85, 101
Mesozoic 1, 5
Metallogenic belts - see also seismotectonomagmatic belts 54
Metamorphic belts - see seismotectonomagmatic belts
Mexico 76
Miami 82
Middle East 76
Midnight 134
Migration of Earth's surficial patterns 45
Minas Gerais 181
Mineral belts, fossil 70
Mineral deposits
evolution 51
evolution during period 0.6 b. y. ago to present 70
evolution during period 3.6 to 0.6 b. y. ago 69
evolution during period 4.6 to 3.6 b. y. ago 66
origin of 51
present distribution 72
Mineral Ridge 132
Minusinsk 154
Miogeosynclines 3, 50, 63
Miracle Mile 181
Mobile belts 1
Mohorivicic discontinuity 74
Mokelumne 85, 101
Monarch 89
Mongollon 87
Mono County 83
Monroe 188
Montana 132
Montezuma 101

Morenci 82
Mormon Basin 120
Morning Star 132
Morobe 101, 106
Morro Velho 181, 196, 197, 202, 208, 216
Mosquito 90
Most favorable areas for future exploration 216
Mother Lode 84, 101, 154, 207, 208
Moto 181, 196
Mount Boppy 132
Mount Isa 133
Mount Morgan 133
Mountain belts 30, 38, 48
Mountain ranges 26
Multiple-fragment accretionary mode 21
Murchison 181
Muruntau 155
Mysore 201
Nacochee 173
National 102
Needle Mountain 133
New Chum 133
New Guinea 103
New Mexico 182
New South Wales 104
New Zealand 87, 104, 212
Neogene 50
Nevada 85, 102, 207, 215
Nevada City 84, 150, 155
Nezhdaninskoye 133
Nicaragua 104, 208, 212
Nome 80, 104, 209
North America 118
North Carolina 155
North Fork 126
North Star 104, 197, 208
Noranda 199, 201, 202
Northair 134
Norilsk 3, 183
Northern Empire 183
Norton Sound 211
Nova Lima 183
Oatman 105, 167
Obduction associated with mineral deposits 65
Oceans 30, 39, 74
Octantal-fragment accretionary mode 20
Octave 167
Ogura 105
O. K. Mountain 134
Okhotsk 6, 79, 105, 156
Old Eureka 105
Odd platforms 4
Olensk 196
Olskaya 79
Omniductive phenomena
description 28, 44, 48
associated with mineral deposits 65

Oquirrh 83
Oregon 134, 211
Original Amador 105
Orogenic belts 47
Orogenic-cratonic structure
of continents
description 36
limit-stop aspects
78
Orogenic deposits 3
Oroville 106
Oroya 183, 197, 198, 200, 208
Otago 106
Ouray 106, 208
Pachuca 106, 208
Paired belts 48
Pakistan 76
Paleozoic 1, 3
"Panta rhei" 195
Park City 107
Parmour 203
Papua New Guinea 75, 106
Patricia 170
Paymaster 203
Peak Hill 183, 189
Pennsylvania 156
Peru 107, 209, 212
Pinos Altos 95
Phanerozoic 73
Philippines 75, 107, 213
Phoenix 146
Placers 85
Plate-tectonics concept
as supplementary tool in exploration for gold 44
causal mechanism 91, 92
continent-continent collisions 134
collisions 91
description 36, 39, 44, 89, 101
questions left unanswered by simple concept 46, 48
Platforms 1, 2, 3
Plymouth 108
Points of model 33, 77
Polar wandering 40
Polarity reversals 39
Polk 188
Porcupine 184, 196, 200, 201, 202,
216
Portland 91, 108
Post-orogenic deposits 3
Post-primordial Earth 22
Post-primordial tectonospheric
accretion 19
Potosi 135
Precambrian - see also Archean
and Proterozoic 47, 74
Pre-orogenic deposits 3
Preston 203
Primary magmas 52
Primordial Earth 13
Prospect Mountain 93
Proterozoic 1, 3, 73

Putintsevskoye 157
Quadrispherical arcs
Aleutians-to-Galapagos 73, 76, 85
Aleutians-to-Gibraltar 76
Bengal-to-Aleutians 73, 76
Bengal-to-Kermadecs 73, 75, 107
Bouvet-to-Bengal 76
Bouvet-to-Kermadecs 77
Galapagos-to-Bouvet 73, 76
Galapagos-to-Gibraltar 76
Gibraltar-to-Bengal 73, 76, 163
Gibraltar-to-Bouvet 77
Kermadecs-to-Aleutians 76
Kermadecs-to-Galapagos 75
Quadruple points 77
Quartz Creek 89
Quartzburg 126
Queensland 136
Radial translatory equilibration within the Earth 15
Ragged Top 159
Rawhide 136
Ray 82
Recent volcanogenic belts 5
Red Lake 170
Reed 146
Republic 108
Rift-ridge system 26, 30
Rhodesia 196, 198, 199, 203
"Ring of fire" 1, 73
Rock Creek 120
Romania 112, 208
Rosita Hills 89
Rotary equilibration within the Earth 13
Round Mountain 108
Rowan 157
Rouyn-Harricanaw 184
Rudisil 158
Rudnyy Altai 157
St. Catherine 158
Salt Lake City 83
San Francisco 81
San Juan 90, 136
San Manuel 82
Sansei 136
Santa Rita 95
Sayan 1, 143
Scandinavia 192
Scarsdale 137
Sea of Japan 76
Sea of Okhotsk 76
Sea-floor spreading 29, 39
Seamount chains 30
Seismic activities 1

Seismic belts - see seismotectonomagmatic belts
Seismicity 37, 48
Seismomagmatic belts - see seismotectonomagmatic belts
Seismotectonic belts - see seismotectonomagmatic belts
Seismotectonomagmatic belt concept
  applied to exploration for gold deposits 217
  description 44, 47, 54, 55, 73, 118, 143, 165, 194
  mineral deposits as interpreted by 48
Seismotectonomagmatic belts
  Archean 194
  Cenozoic 73
  description 44, 47
  fossil 55
  Mesozoic 118
  oldest 54
  Paleozoic 143
  Proterozoic 165
Seneca 137
Serra de Jacobina 184
Seward Peninsula 104, 211
Shamvaian 53
Sheep 151
Sheep Creek Camp 137
Shenandoah 109
Shturmanskoye 137
Siberia 1, 3, 4, 143, 192, 195, 196, 216
Sierra County 158
Sierra Nevada 84, 209, 215
Silver Peak 138
Silverton 109, 208
Simon 151
Siscoe 185
Skaergaard 139
Skellefte 168
Slave 195
Snelling 109
South America 28, 76
South Carolina 138
South China Sea 76
South Dakota 185
South Pacific 46
South Pass 191
South Verkhoyan 138
South Viet Nam 75
Spain 76
Sparta 120
Spreading lineaments 46
Squaw Creek 159, 186
Sredniy 159
Sri Lanka 75
Staryy Oskol 187
Steamboat Springs 139

Steeple Rock 95
Stratton 91, 109
Strawberry 171
Stress release within Earth
29
Structures associated with
gold deposits 2
Sturgeon River 187, 203
Subduction associated with
mineral deposits 65
Subtectonosphere 10
Subtectonospheric equator 33
Subtectonospheric fracture
system 33
Sudbury 3, 187
Sumatra 75
Summitville 89
Sumpter 120
Sunnyside 109
Superior 82, 195
Surficial features 36, 73
Surficial manifestations of
the Earth's internal
behavior 30
Surficial patterns 30
Susanville 126
Sweden 168
Sylvanite 200
Taiga 100
Takatama 109
Tarkwa 2, 187, 210, 216
Tarnagalla 138
Tayoltita 110
Teck-Hughes 200
Tectonic activity
associated with intrusive-
extrusive activity 62
description 1, 5.
Tectonic belts - see seismotec-
tonomagmatic belts
Tectonomagmatic belts - see seismo-
tectonomagmatic belts
Tectonomagmatic episodes 47
Tectonospheric Earth Model
as supplementary tool in the
exploration for mineral
deposits 35, 214
basic points of 32
coordinate system defined by
33
driving mechanism for 13, 30,
36
Earth's internal behavior in-
terpreted by 30
description of 9, 10
evolution of mineral deposits
interpreted by 66
global surficial manifesta-
tions of 30
intrusive and extrusive acti-
vity interpreted by 53
poles of 33
salinet implications of 23
Telluride 139, 208

Tennessee 188
Tertiary placers 215
Thailand 75, 211
Thames 110, 208
Thrusts associated with intrusive-extrusive activity 63
Thunder Bay 188
Tim-Yastrebovka 187
Timmins 199, 203
Tincup 89
Tintic 110
Titagara 160
Toburn 200
Tombstone 82, 167
Tonga-Kermadecs 76
Tonopah 111, 208
Transbaikal 140, 159, 160
Transcarpathia 112
Transform faults 48
Transvaalian 53
Transverse translatory equilibration within the Earth 15
Transylvania 112, 208
Treadwell 112, 208
Triple junction - see quadruple points
Toulumne 85
Turquoise 82
Tuscarora 113
Two Bit 171, 186
Types of deposits 1
Tympie 136
Ultramafic rocks 48
Unbalanced conditions within the Earth 27
Unga 119
Union 113
Union Pass 167
United States 143, 165, 192, 212
Uplifts associated with intrusive-extrusive activity 63
Upper Applegate 128
Upper Burnt River 120
Upper-mantle evolution 56
Urals 1, 4, 160
USSR 210
USSR lodes 160
USSR placers 113, 161
Utah 113
Val d'Or 203
Vasterbotten 168
Verkhoyansk 3, 118
Vertical motion 75
Victoria 113, 161
Vindicator 91, 113
Virginia 162
Virginia City (Montana) 113, 209
Virginia City (Nevada) 114, 208
Virtue 120
Vitim 178
Volcanic belts - see also seismotectonomagmatic belts 26

Volcanogenic belts - see also
seismotectonomagmatic
belts 8
Voronezh 187
Vulture 82
Waihi 208
Wallapai 82
Warren 140
Washington 114
Wave guides 23
Weaver 82
Wenatchee 115
West Belt 84
West Irian 75
Western Australia 188
White 173
Wildman 100
Willow Creek 80
Wilson 115
"Wire gold" 170
Witwatersrand 2, 190, 208,
210, 216
Woods Point 141
Wright-Hargreaves 200
Wyalong 141
Wyoming 115, 191, 209, 215
Yakataga 119
Yakut 115, 118, 163
Yalu 136
Yana 137
Yankolyma 3, 141
Yatani 115
Yelete 196
Yellowknife 170, 204
Yenisei 163, 192, 209
Young placers 208
Yilgam 188, 192
Yuba River 85, 117
Zaire 181, 196, 204
Zolotaya 163